SpringerBriefs in Well-Being and Quality of Life Research

More information about this series at http://www.springer.com/series/10150

Gonçalo Nuno Figueiredo Dias
Micael Santos Couceiro

Active Ageing and Physical Activity

Guidelines, Functional Exercises
and Recommendations

 Springer

Gonçalo Nuno Figueiredo Dias
Faculty of Sport Sciences and Physical
  Education (CIDAF)
University of Coimbra
Coimbra
Portugal

Micael Santos Couceiro
Ingeniarius, Ltd.
Coimbra
Portugal

ISSN 2211-7644　　　　　ISSN 2211-7652　(electronic)
SpringerBriefs in Well-Being and Quality of Life Research
ISBN 978-3-319-52062-9　　　ISBN 978-3-319-52063-6　(eBook)
DOI 10.1007/978-3-319-52063-6

Library of Congress Control Number: 2016962051

© The Author(s) 2017
This work is subject to copyright. All rights are reserved by the Publisher, whether the whole or part of the material is concerned, specifically the rights of translation, reprinting, reuse of illustrations, recitation, broadcasting, reproduction on microfilms or in any other physical way, and transmission or information storage and retrieval, electronic adaptation, computer software, or by similar or dissimilar methodology now known or hereafter developed.
The use of general descriptive names, registered names, trademarks, service marks, etc. in this publication does not imply, even in the absence of a specific statement, that such names are exempt from the relevant protective laws and regulations and therefore free for general use.
The publisher, the authors and the editors are safe to assume that the advice and information in this book are believed to be true and accurate at the date of publication. Neither the publisher nor the authors or the editors give a warranty, express or implied, with respect to the material contained herein or for any errors or omissions that may have been made. The publisher remains neutral with regard to jurisdictional claims in published maps and institutional affiliations.

Printed on acid-free paper

This Springer imprint is published by Springer Nature
The registered company is Springer International Publishing AG
The registered company address is: Gewerbestrasse 11, 6330 Cham, Switzerland

Foreword

"Active Ageing and Physical Activity: Guidelines, Functional Exercises and Recommendations"

Globally, the 65+ segment of the population is rapidly increasing. Medical and technological advances over the past century have had a significant and positive impact on the health and overall well-being of our nations' citizens and longer life expectancies.

Regular participation in physical activity (PA) is essential for maintaining good health, particularly as we age. Considerable research evidence has demonstrated that individuals who are active in their older adult years exhibit lower rates of disease (e.g. cardiovascular disease, diabetes, colon and breast cancer) and disability. In contrast to their physically inactive counterparts, physically active older adults maintain healthier body weights, higher cognitive function and better overall functional health.

Despite the wealth of research evidence that has identified physical inactivity as a key risk factor for a number of chronic medical conditions that result in premature disability and/or mortality in the older adult years, a large proportion of older adults (≥65 years) do not currently meet global physical activity recommendations. Older adults ageing with a disability are even less likely to engage in any type of leisure time activity when compared to older adults without disability. This is a particularly troublesome finding given that individuals with disability who regularly engage in physical activity derive similar health benefits. A reduction in the level of functional impairment and improvements in perceived quality of life have also been demonstrated to individuals with disability who participate in physical activity, even at lower levels of intensity.

"*Active Ageing and Physical Activity: Guidelines, Functional Exercises and Recommendations*" promises to be an exceptional resource for any professional working with the older adult population. Drs. Dias and Couceiro address the role of physical activity in promoting active ageing from a multidisciplinary perspective, combining essential theoretical knowledge of the ageing process and common age-associated diseases with practical information needed to design age-appropriate physical activity programmes that can be safely and effectively implemented by trained professionals.

I look forward to adding this contemporary resource to my professional library and sharing the theoretical and practical knowledge presented in this book with undergraduate and graduate students currently preparing for professional careers working with older adults in rehabilitation and physical activity settings.

Debra J. Rose, Ph.D., FNAK
Director, Center for Successful Aging
California State University
Fullerton

Acknowledgements

The authors would like to acknowledge Dr. Inês Carvalho, Dr. Pedro Celaviza, Dr. Pedro Mendes, Dr. Paula Rodrigues, Dr. Pedro Martins, Dr. João Ventura, Dr. Diana Rodrigues, Dr. Fernanda Dias, Dr. Maria de Lurdes Almeida, Dr. Jorge Felício, Dr. Cristina Leandro, Dr. Maria António Castro, Dr. José Pacheco, Dr. António Leal, Dr. Luísa Almeida, Dr. Filipa Sousa, Ms. Ascensão Neves, Ms. Maria Dolores Batista, Mr. Horácio Queiroz, Mr. Frederico Almeida and, finally, the Coimbra School of Education (Escola Superior de Educação, Instituto Politécnico de Coimbra), for all the technical support.

This research was supported by the Portuguese Foundation for Science and Technology (FCT) under the Grant SFRH/BPD/99655/2014.

Contents

1 **Introduction: New Paradigms of Active Ageing** 1
  Gonçalo Nuno Figueiredo Dias, Micael Santos Couceiro,
  Polybio Serra e Silva, Maria António Castro, Maria Aurora
  Branquinho, Rui Mendes and Inês Cláudia Rijo de Carvalho
  1.1 Population Ageing 2
  1.2 Active Ageing: Retrospective and Future Trends 7
  1.3 Physical Activity in the Active Ageing Process 10
  1.4 Disorders and Psychomotor Rehabilitation 11
       1.4.1 Cerebrovascular Accidents 12
       1.4.2 Heart Disease 13
       1.4.3 Pulmonary Disease 14
       1.4.4 Osteoporosis 14
       1.4.5 Metabolic Disorders (Diabetes) 15
  1.5 Conclusions and Practical Implications..................... 15
  References... 16

2 **Physical Activity Benefits in Active Ageing** 21
  Gonçalo Nuno Figueiredo Dias, Micael Santos Couceiro,
  Pedro Mendes and Maria de Lurdes Almeida
  2.1 Background 21
  2.2 Morphological and Functional Changes of the Elderly 22
  2.3 Body Composition 23
  2.4 Cardio-respiratory Capacity 23
  2.5 Musculoskeletal System 24
  2.6 Central Nervous System................................... 24
  2.7 Sensory and Perceptive System 25
  2.8 Movement Duration and Motor Reaction 26

		2.9	Health Benefits of Physical Activity	27
			2.9.1 Aerobic Capacity	27
			2.9.2 Muscular Strength	28
			2.9.3 Flexibility	29
			2.9.4 Balance	29
			2.9.5 Biopsychosocial Model	30
		2.10	Conclusions and Practical Implications	31
		References		31
3	**Activity Programmes for the Elderly**			35
	Gonçalo Nuno Figueiredo Dias, Micael Santos Couceiro and Rui Mendes			
		3.1	Regular Physical Activity and Healthy Ageing	35
		3.2	Physical Fitness Evaluation	38
		3.3	Physical Activity Prescription	39
			3.3.1 Structure	39
			3.3.2 Frequency	39
			3.3.3 Duration	40
			3.3.4 Intensity	40
			3.3.5 Repetitions per Exercise	40
			3.3.6 Technical Indications	40
		3.4	General Exercises	40
			3.4.1 Stick	41
			3.4.2 Ball	45
			3.4.3 Hoop	50
			3.4.4 Resistance Band (Part 1)	56
			3.4.5 Resistance Band (Part 2)	61
		3.5	Strength Exercises	74
			3.5.1 Dumb-bells	75
			3.5.2 Neoprene Ankle Weights	81
		3.6	Partner Exercises	86
			3.6.1 Body Strengthening	86
			3.6.2 Body Language	93
		3.7	Return to Resting State	98
		3.8	Conclusions and Practical Implications	102
		References		102
4	**Technology for the Active Senior**			105
	Micael Santos Couceiro and Gonçalo Nuno Figueiredo Dias			
		4.1	Mixed Reality Serious Games and Robotics	105
		4.2	Mixed Reality Serious Games	107
			4.2.1 Serious Games	109
			4.2.2 Mixed Reality	109
			4.2.3 Wearable Technology	110

	4.3	Robotics		111
		4.3.1	Appearance and Physical Characteristics	112
		4.3.2	Real-Time Assistance and Monitoring Over the Internet	113
		4.3.3	Autonomous Navigation and Operation Under Dynamic Environments	114
	4.4	Conclusions and Practical Implications		115
	References			116
5	**Conclusions**			119
	Gonçalo Nuno Figueiredo Dias and Micael Santos Couceiro			
	5.1	Conclusions		119
	5.2	Practical Implications		120
	5.3	Recommendations		120

About the Authors

Gonçalo Dias *Faculty of Sport Sciences and Physical Education (CIDAF), University of Coimbra, Portugal Mailing address: Estádio Universitário de Coimbra, Pavilhão 3, 3040-156, Coimbra, Portugal, e-mail address: goncalodias@fcdef.uc.pt*

Gonçalo Dias obtained both M.Sc. and Ph.D. degrees on sport sciences at the Faculty of Sport Sciences and Physical Education of University of Coimbra and the B.Sc. degree at the Coimbra College of Education from the Coimbra Polytechnic Institute. He is currently a professor of advanced topics in motor learning at the Faculty of Sport Sciences and Physical Education of University of Coimbra (Master Course in Youth Sports Training) and a member of the Scientific Committee within the same institution. Over the past 10 years, he has been conducting scientific research on several areas associated with sports and health, namely physical activity among older adults, motor control and sport sciences. Besides research and lecturing, he has been in charge of the organization of events, both in the public and private domains, as coordinator of the County Sports Office from Arganil, Portugal. Additionally, he is responsible for the programme "Physical activity to older adults", which brings together more than 100 elderly aged between 65 and 100 years old. This activity resulted in the publication of three books and several articles that portray the work of his research in 20 private institutions of social solidarity in Portugal, being currently considered one of the Portugal's leading experts in this field.

Micael S. Couceiro *Ingeniarius, Lda.*
Artificial Perception for Intelligent Systems and Robotics (AP4ISR), Institute of Systems and Robotics (ISR), University of Coimbra
Mailing address: Rua da Vacaria, n.37, 3050-381, Mealhada, Portugal, e-mail address: micael@ingeniarius.pt

Micael S. Couceiro obtained the B.Sc., Teaching Licensure, and master's degrees on electrical engineering (automation and communications), at the Coimbra School of Engineering (ISEC), Coimbra Polytechnic Institute (IPC). He obtained the Ph.D. degree on electrical and computer engineering (automation and robotics) at the Faculty of Sciences and Technology of University of Coimbra (FCTUC), under a Ph.D. grant from the Portuguese Foundation for Science and Technology. Over the past 6 years, he has been conducting scientific research on several areas, namely robotics, computer vision and sports engineering, all at the Institute of Systems and Robotics (ISR-FCTUC). This resulted in more than 25 scientific articles in international impact factor journals and more than 50 scientific articles at international conferences. Besides being currently a researcher at ISR, he has been invited for lecturing, tutoring and organization of events (e.g. professional courses and national and international conferences), both in the public and private domains. He is a co-founder and currently the CEO of Ingeniarius—a company devoted to the development of smart devices, from which performance analysis in sports has been one of its main markets.

Chapter 1
Introduction: New Paradigms of Active Ageing

Gonçalo Nuno Figueiredo Dias, Micael Santos Couceiro, Polybio Serra e Silva, Maria António Castro, Maria Aurora Branquinho, Rui Mendes and Inês Cláudia Rijo de Carvalho

Abstract The main purpose of this chapter is to describe active ageing as a continuous and unavoidable process. An integrated and holistic approach is needed, which highlights the advantages of psychomotricity and gerontomotricity. Such an approach can decrease physical and social isolation in the elderly, and even help them to find their affordances. Therefore, their opportunities for socialization and recreation can be increased. Elderly quality of life stretches beyond the physical and biological dimensions. Healthy ageing presupposes a dynamic balance between body, cognition and affection. The elderly think, feel and move differently. Therefore, they require special care concerning their physical activity and the management of the organic, nutritional and physiological aspects that affect their ageing bodies. From this perspective, the topic of active ageing comprehends healthy lifestyles and physical activity. These recommendations, which are widely known in Western societies, aim to prevent disease and promote health. In addition, active ageing, in the broad sense, should assume a paradigm shift that adequately responds to aspects related with the increase in longevity, quality of life and health among the older people. According to the state of the art, physical activity can play a crucial role in the protection against age-related morbidity and in the increase of longevity. Regardless of the age when physical activity starts, changes in sedentary patterns, even among those older than 85, can substantially reduce mortality and functional disability. The adaptations introduced in the movement, if performed adequately, may contribute to improve not only individuals' health, but also their quality of life. Finally, physical activity may effectively improve ability by improving several functions of the body, such as strength, flexibility, resistance and general physical aptitude. However, it is necessary to adapt physical activity recommendations to older people, in order to cater for their specific needs. It is also essential to use several types of exercise which can correct or improve the functional limitations identified.

Keywords Active ageing · Physical activity · Quality of life · Health

---

Areas of expertise: Tourism Management and Planning Modern languages, Literature and Linguistics Gender studies

1.1 Population Ageing

Population ageing is a global phenomenon which affects all developed countries. It reflects the improvement in global health, and is thus the outcome of an unquestionable triumph of modern society. However, this phenomenon is also marked by deep-seated inequalities, as those revealed by indicators such as life expectancy at birth. For example, while the population in Japan has the longest life expectancy in the world at 82.2, at the other end of the scale, life expectancy is approximately half of that in several African countries (Dias et al. 2015). Health costs raise as population ages, particularly in the last two years of life, regardless of age. Since people now live longer than before, it is important to guarantee that their additional years of life are healthy, so that health costs are sustainable. Recent studies have demonstrated that the elderly population has increased significantly over the last years (e.g., Dias et al. 2015). In fact, the World Health Organisation (WHO) has predicted that the elderly will exceed 780 million people in 2025. Moreover, according to Eurostat forecasts for Portugal, the percentage of older people in the Portuguese population in 2040 will reach 20.6% (Dias et al. 2014).

The United Nations report (*World Population Ageing*: 1950–2050) has revealed that those who were 60 or older in 1950 corresponded to 8% of the population. It has also predicted that this age group will comprehend 21% of the population by 2050. This report indicates that, in 2002, the number of people aged 60 or older was 629 million, and that it will reach 2000 million by 2050 (Nations 2001; Cruz 2009; Cabral et al. 2013). Moreover, this report shows that the ageing population will increase sharply in the next 25 years. As a result, it is predicted that the elderly (80 years or more) will have the fastest growth rate, and that it will correspond in the future to 12% of the total number of people aged 60 or older.

The pace of population ageing is faster in developed countries than in developing countries (Nations 2001; Oliveira 2008; Cruz 2009; Cabral et al. 2013). Still according to the United Nations report, the average age of the world population in 2002 was 26. Considering the data for the present year, the youngest country was Yemen where the average age was 15, and the oldest one was Japan, where the average age was 41. This way, it is expected that, by 2050, the average age increases ten years, i.e. to 36 years. Therefore, it is predicted that in 2050 the youngest country will be Niger, with an average age of 20. Furthermore, Spain will have the oldest population in Europe, with an average age of 55. Finally, it is estimated that Portugal will be ranked as the sixth oldest country in the world and the fifth in Europe (Nations 2001; Stella 2008; Cruz 2009; Cabral et al. 2013).

Concerning the ageing index, Asia will undergo the greatest increase in older people until 2050. In Africa, the increase in the proportion of the population aged 60 or older will correspond to 310% in the same period. In contrast, the proportion of older population will decrease (Bowling 2007; Cruz 2009; Cabral et al. 2013).

From this perspective, several global organizations, such as the United Nations (UN), the World Health Organisation (WHO) and the Organisation for Economic Cooperation and Development (OECD), have indicated significant changes in terms of demographic ageing and age composition of the population. This way, between 1970 and 2025, it is predicted that the older population will increase by 223%. Against this background, there will be about 1.2 billion people aged 60 or older in 2025 and about 2 billion in 2050, and 80% of this population will be concentrated in developed countries (Cruz 2009; Magalhães 2005; Cabral et al. 2013).

Older adults are now the fastest growing age group, particularly those over the age of 85, whose growth rate has been estimated at 4% per year. This increase will be particularly noticeable among women. At the same time, life expectancy at age 60 will increase by about 22 years until 2050. With regard to the European scenario, it may increase by about 5 years until 2050, and this growth will be most noticeable in the age groups of 80–90 years (Magalhães 2005; Stella 2008; Ageing Report 2012).

The increase in ageing population emerges as the main demographic trend occurring in in Portugal in the most recent years. Considering the Portuguese socio-demographic situation, it is expected that the tendency of a constrictive age pyramid will be accentuated by 2050, with the proportion of older population corresponding to 35.72% and that of children and youth corresponding to 14.4%, and life expectancy estimated at 81 years (Magalhães 2005; Ageing Report 2012).

In 2011 Portugal registered a longevity index of 79.20 years of age, namely 80.57 for women and 74.0 for men. The projections for 2050 estimate a significant increase in this index, which predicts that people will live on average 81 (Ageing Report 2012; Cabral et al. 2013). Data published by the Department of Economic and Social Affairs[1] indicated the existence of 300 people aged 100 or over in Portugal, and it is predicted that this figure will reach 1800 in 2025 and 6400 in 2050. Moreover, women have been the majority (58%) in the age group above 65, which is the result of a trend referred to as the "feminisation of ageing", which has been increasingly observed in the Portuguese society since 1900 (Cabral et al. 2013).

Demographic ageing has been one of the most relevant phenomenons occurring in developed societies in the twenty first century. However, population ageing should not be regarded as a burden, but rather as a challenge and an opportunity. Therefore, it is crucial to invest in measures for social inclusion, for the integration of the elderly, so that the additional years of life are healthy and fulfilling years as a biological phenomenon, ageing is associated with increasing difficulties in maintaining

---

[1]ONU—United Nations (1998). Department of Economic and Social Affairs: World Population Projections to 2150.
ONU—United Nations (2001). Department of Economic and Social Affairs: World Population Prospects: The 2000 Revision.

homeostatic balance, which becomes progressively more vulnerable with age. For example, the reproductive capacity of cells decreases after 20 years of age.

The biological changes inherent to the natural ageing process are reflected in all organic systems, as well as in biological and sleep rhythms. Many theories have been proposed, which can be grouped into two broad categories: a biological and a psychosocial one. However, none of these theories in isolation is sufficient to provide a satisfactory explanation.

In addition to the physiological changes, ageing is accompanied by psychological and also social changes, which may be experienced within the family (widowhood), at work (retirement or leisure occupations), in social relations and routines, which can have consequences for the reputation of the elderly. Our social/genetic background and our life stories greatly influence the way we adapt to the ageing process. Psychological changes concern the behavioural capabilities of the individual to adapt to the environment, such as intelligence, learning ability, feelings and emotions. In the case of the elderly, these psychological changes may pose obstacles to their adaptation to new social roles, or result in lack of motivation, difficulty in planning the future or adapting to rapid changes, lead to low self-esteem and low self-image, as well as a need to work on emotional and social losses. The most optimist and healthy individuals adapt more easily to the changes resulting from ageing. These changes are not always losses, but there may also be gains, e.g. in terms of intelligence and learning capabilities, which can progress as long as they are exercised (Serra e Silva et al. 2015).

Activity is crucial for healthy ageing. Therefore, the individuals that perform social tasks and activities are the ones who live more happily throughout their years. In that sense, a meta-theory conceived ageing as social, psychological and biological processes, and considered development and ageing to be synonyms. There is also another theory that advocates the existence of a balance between genetics and the environment, which are the aspects that influence the ageing process (Serra e Silva 2012). More recently, two theories have been developed. One of them considers that satisfaction with life is attainable if individuals focus on a transcendent dimension at a cosmic level, from a perspective that leads to the acceptance of the mystery of life. The other theory advocates regards ageing in parallel with general theories and chaos theory, since human functioning is seen as dynamic and as finding opportunities to transform itself in chaos. Hence, three patterns emerge: mortality and quality of life (Serra e Silva et al. 2015).

There are multiple myths and stereotypes concerning ageing. Our belief is that they are the outcome of ignorance about the ageing phenomenon, which leads to false perceptions that end up associating this natural process to sickness, boredom, selfishness, dependence and loss of social status. These preconceived ideas may cause social isolation among the elderly, leading to pathological ageing, since society ends up discriminating and stereotyping older people. Myths are complex and unconscious, implying erroneous interpretations and symbolic representations, which are essentially negative and discriminatory. This type of discrimination and stereotyping against individuals on the basis of their age, which is particularly experienced by seniors, is called ageism (Serra e Silva et al. 2015). These myths and stereotypes against older people concern aspects such as the chronological process itself, lack of productivity, senility, loss of sex drive, lack of serenity, deterioration of intelligence, disengagement with the future, isolation and alienation. They are verbalized in popular sayings, such as "you can't teach an old dog new tricks". However, such preconceived ideas do not necessarily reflect the reality, since the last stage of an individual's life can be as rich and heterogeneous as previous stages (Serra e Silva et al. 2015).

There are two main representations of the ageing process. One of them is negative, and regards it as associated to poverty, social isolation, loneliness, sickness and dependency. The other is positive, and it is related with aspects such as freedom, economic stability and availability for leisure and culture. Ageing is more often than not erroneously associated with disease, as if they were synonyms. In fact, many older people are strong and healthy, although with some level of decline, which is part of the normal ageing process. However, some point out the existence of delirium, dementia and depressive symptoms, which has not been proved. In fact, older people seem to be more resilient to life adversities. This protects them against depressive symptoms (Serra e Silva et al. 2015).

Although there are gains until the age of 40, which is followed by a period of stability, declines start at the age of 60. However, this process depends on several factors which vary individually. Hence, it is not correct to systematically associate lack of productivity and old age, since many people discover new talents or interests in this life stage. Although fluid intelligence, i.e. the processing speed, may slowly decline after age 20, crystallized intelligence remains intact. With age, older people become wiser, since wisdom can only be acquired with experience (Serra e Silva et al. 2015). The loss of sexual desire is another myth, since older people have their sexuality adapted to their conditions and life experiences. Sexuality is not restricted to intercourse itself, but it is related to the way men and women gesticulate, walk, dress, their life and thinking patterns, which does not mean that sexuality does not exist in old age, but that, on the contrary, older people may remain sexually active and interested in sex (Serra e Silva et al. 2015).

It is true that we can find older people with anxieties and conflicts, but also with a true sense of serenity and well-being. Still, one of the myths is related with the idea of youth as a life perspective and old age as a death perspective. In addition, disengagement with the future is not the norm. In fact, many older people are interested in learning and obtaining new knowledge. An example is older individuals' use of computers as a communication tool and source of knowledge. Another myth is that older people are afraid to die. However, a healthy desire to be alive should not be confused with anxicty towards death. Moreover, it is not possible to claim that this anxiety increases with age, since it is middle-aged people who fear death the most because of their greater responsibilities. They are called the Sandwich Generation, which is a generation of people who care for their ageing parents while supporting their own children (Serra e Silva et al. 2015). In this phase of life, many older people feel personal fulfilment and are more familiarized with death than any other age group.

Fourteen stereotypes in relation to the elderly have been observed, which reveal ignorance towards ageing, as already observed. Therefore, it is crucial to debunk these myths and preconceived ideas, so that ageing is understood and perceived as a natural life phase, with both losses and gains. Moreover, the elderly population comprises a heterogeneous group, where each person is unique. It is, therefore, necessary to adopt a different attitude as to how society perceives ageing, as this affects how the elderly perceive themselves. Put it differently, we should not accept that the elderly consider themselves as disabled, sick and unnecessary but, on the contrary, we must demand from them to get adapted to this new stage of life, rediscovering themselves, finding new goals, new ways of leisure, increasing their social network and sense of self-worth (Serra and Silva et al. 2015).

In order to increase the lifespan and successful ageing, one has to consider that ageing will always be seen in different ways, either negatively, as a period of vulnerability and greater reliance, or positively, as a way to achieve serenity and wisdom. It is natural that tobacco addiction, diseases, low environmental stimulation, limited and fragile social networking, less healthy diet and physical inactivity, among others, may result in a gradual loss of the ability to adapt. Contrariwise, a life filled of positive experiences, obstacles properly surpassed, systematically ongoing activities, continuous physical exercise, enthusiasm and dedication and so on can provide a more positive ageing. All these aspects are related to how individuals grow older and how they perceive their own ageing (Serra and Silva et al. 2015).

Successful ageing is an adaptation mechanism to the specific conditions of old age and the environment, depending not only, but mainly, upon the way the person lived his/her life. The successful ageing definition still lacks some consensus in the literature. However, there are many strategies for successful ageing that intends to reduce the prevalence of morbidity and prolong the well-being of the elderly. Yet, there needs to be an early intervention to foster preventive behaviours instead of acting only when facing disease and disability. Strategies for successful ageing need to be preventive or corrective, or both, depending on what individuals want and the available resources (Serra and Silva et al. 2015).

The successful ageing concept was presented in 1994, being the term used for the purpose of describing the existence of ageing without or with only minor loss of functions, fighting back the negative idea, and not always true, that ageing is inevitably related with the *4 Ds*: *Dependency, Disease, Disability* and *Depression*. There is then a boundary between normal ageing and successful ageing, in which individuals have to provide high levels of capability to operate in multiple dimensions, emphasizing the importance that the elderly can attain a certain potential, which is individual to each one of them. Older people should feel good about themselves, and this should be seen in a physical, psychological or social dimension, which, inevitably, will also allow them to feel good about others. Successful ageing is then a combination of several factors, such as longevity, happiness and lack of disability. It seems indisputable that, if there is any need for the elderly to adapt to the outside conditions, then the society also has to find a proper fit to each individual. Put it differently, the society should draw the attention to the social and political issues involved, such as their integration in the labour market, stimulation of social relations, and the sharing of responsibilities (Serra and Silva et al. 2015).

The society has been witnessing an increased population ageing as a result, not only, of the increased longevity. But, in a reality wherein one can live longer, adding years of life may not be as relevant as adding life to these years. Quality of life is now a concept of utmost importance, as more and more we realize that life is not merely a problem of subsistence, but should also embrace value and dignity. The promotion of health has become a primary objective of healthcare systems, alerting the public to the need to adopt, in all its dimensions, a healthy lifestyle. Unfortunately, the definition of quality of life is not that consolidated in the literature, since it is characterized by being very complex and covering various aspects of life. Still, there are several authors who have studied this topic in depth, especially in an elderly context. The WHO has developed an attempt to assess the quality of life, which takes into account both physical and psychological domains, the level of independence, the environment and spirituality, religion and personal beliefs. Many works have been focusing in the welfare notion, addressing it in a mental health, social psychology and gerontology perspectives, recognizing it as a multidimensional phenomenon that comprises optimism, happiness, personal determination, positive and negative emotions, and others (Serra and Silva et al. 2015).

1.2 Active Ageing: Retrospective and Future Trends

The OECD (Organisation for Economic Cooperation and Development) defines active ageing as the ability of people who grow older to lead a productive life in society and the economy (Dias, Mendes, Serra e Silva, & Banquinho, 2014, 2015). This means that people are able to determine how they should split their lifetime between learning, work, leisure and care (for others) activities. Adducing other contributions to this definition, the WHO (2001 and 2005) argues that active ageing covers individual health opportunities, participation and security, with the main objective to increase the quality of life during old age. For this organization, the active ageing, as an existing paradigm, should cover all persons, including those who are frail, physically disabled and under ongoing medical care (Neda et al. 2011; Cabral et al. 2013). In spite of this, active ageing should be seen as a continuous and inevitable process, being a stage of life where one needs to understand the many changes occurring in the old age. As such, a unifying and integrated approach that encompasses both psychomotor and gerontomotricity advantages can reduce the physical and social isolation of the elderly, leading him/her to discover new possibilities for action (i.e. affordances), thereby increasing his/her input in socialization and recreational activities, also reducing the vulnerability to possible diseases and disorders (Lee et al. 2008).

The quality of life of the elderly then goes far beyond the physical and biological dimensions. In fact, it is within this environment of social integration wherein the elderly may have the opportunity to improve their quality of life and, at the same time, be integrated as key players in our society. Healthy and active ageing requires a dynamic balance between body, cognition and affectivity. Thus, as the elderly thinks, feels and moves differently, a special care is required in terms of body practices and management of organic, nutritional and physiological aspects.

Nevertheless, in most approaches designed for the third age, physical activity and physical health education emphasize, mainly, in the physical component of the "body-object". This perspective of the body as something only physical can be restrictive for the elderly as he/she should be perceived as a bio-psycho-social being (Zijlstra et al. 2007; Dias et al. 2014). An important indicator of active ageing refers to the degree of functionality that can be measured while performing basic and instrumental activities of daily living. Hence, an effective psychomotor intervention can stimulate a dynamic balance between body and mind. Likewise, this approach may also promote the full development of the elderly in a transdisciplinary and integrative perspective.

Bearing this in mind, ageing has a multifactorial underlying matrix, where the elderly emerge as a bio-psycho-social being with different needs and capacities. In this respect, several factors are associated with the ageing process, such as molecular, cellular, systemic, behavioural, cognitive and social, among others (Spirduso 2005; Howell et al. 2008; Dias et al. 2014). As previously stated, according to the WHO, there is a set cultural, social and behavioural order factors which must be weighed in the active life of the elderly (Capucha 2005; Magalhães 2005; Cabral et al. 2013), namely,

1. Culture and gender—the culture that surrounds the life of the elderly can influence the whole ageing process. It is also necessary to take into account the issues related to gender, since ageing as a "man" is not the same than as "woman";
2. Health systems and social services—all elderly should have equal access to health systems and social services. It is then imperative to follow a health strategy that includes measures to promote, prevent and provide continuous care;
3. Behavioural factors—society, in general, must adopt a pedagogical and integrative behaviour towards the elderly. The life experience of older people should be valued at several levels;
4. Physical environment—environmental conditions where the elderly live may influence the ageing process. Note that a poor home, or the existence of architectural barriers, can influence the lives of the elderly, especially those facing reduced mobility;
5. Social environment—the social environment, generated by the family and the general society, as well as the access to education and learning throughout life, are key elements in promoting the independence of older people and further developing their skills;

6. Economic factors—according to the WHO (2001, 2005), the elderly population is very vulnerable to poverty, not only due to low income (e.g. pensions), but also due to labour market exclusion after entering the retirement age. It is, therefore, important to recognize and revitalize the important role that the elderly have within their family economy and their contribution to society in general.

Active ageing aims at healthy lifestyles within the elderly population, whose recommendations are well known in western societies, focusing in the prevention of diseases. However, despite the healthy compromise between active ageing, quality of life and health, we note that authors, like Cabral and Silva (2010), argue that healthy habits in the old age are like a sort of replacement for social inequalities in health. For the same authors (2010), these healthy habits assume, essentially, a preventive role that public policies and health professionals should be promoting. Hence, active ageing should go through a paradigm shift to adequately tackle the issues related to the increasing longevity, quality of life and health of the elderly. For instance, the increased longevity phenomenon is especially observed in developing countries, in such a way that, according to the WHO (2001, 2005), the overall ageing during the upcoming decade will cause a significant increase in the demand for social and economic responses worldwide. On this basis, the current debate on ageing covers, among other issues, health policies and the role of the family in social security systems (Paúl 2005). In addition, the increasing prevalence of chronic and disabling diseases are on the basis of a paradigm shift in health, considerably constraining the degree of preservation of functional capacity (Dias, Mendes, Serra e Silva, & Banquinho, 2014, 2015).

Healthy ageing encompasses a broad meaning; a multidimensional interaction between physical health, mental health, family support, economic independence and social integration (Fernández-Ballesteros 2002; Fernández-Ballesteros et al. 2010). As such, public health challenges go from mechanisms to act upon the physical level to both mental and social well-being (Fonseca 2005; Neda et al. 2011). For instance, the literature highlights that the sociological profile of the elderly in Portugal is strongly characterized by the low income, high illiteracy, precarious housing conditions and high prevalence of chronic diseases, social isolation and diminished occupation. Lima (1999) emphasizes that the old age is directly related with the disease, widowhood and death, seeing this life stage as unfavourable, often dramatic and unwanted. That being said, Lima (1999) also notes that a new sociological paradigm has been emerging, where more and more older people play sports, dance and carry out other activities that are normally associated with the younger population. According to the same author, this increased social participation of older people cannot be solely explained by the increased ageing population, but yet by a set of social, sociological and professional phenomena. On the other hand, this increased social participation of the elderly population, leading to a stronger inter-generational relationship, is one of the WHO guidelines towards the implementation of "healthy cities" (Neda et al. 2011).

1.3 Physical Activity in the Active Ageing Process

Due to the continuous advances in science, it has been possible to increase the life expectancy of individuals. Therefore, the current expectation is that both men and women may live, easily, more than three-quarters of a century. Naturally, an aged human body will also lead to a higher level of detrition and loss of functional capacity. The latter, i.e. the increased challenge to independently perform basic and instrumental daily life activities, can lead the elderly to a suffering state, exacerbation of the disease, frailty and vulnerability. Yet, all this can be prevented with physical activity maintenance, as long as it is adequately adapted to each individual's physical, social, mental and cognitive conditions (Castro 2015).

According to the research carried out so far (King and Guralnik 2010), maintaining a periodic physical activity programme allows to build a protective layer against age-related morbidities, being considered as an enhancer aspect of increased longevity. Regardless of when the physical activity is initiated, the changes in sedentary patterns, even in older individuals above 85 years (Keysor 2003; Pahor et al. 2006; Lobo et al. 2011), are able to substantially reduce mortality and disability. Remarkable, many studies show that individuals with more than 85 years old following physical activity programmes live, in average, 3 years more (Stessman et al. 2009). However, in cases in which the elderly might suffer from reduced mobility, it becomes necessary to adequately prepare an adequate physical activity programme. The movements should be contextually adjusted with the intent to contribute towards improving the health of individuals, as well as their quality of life, with the intent to increase their longevity (Castro 2015).

The International Classification of Functioning, Disability and Health (ICF) proposed by the WHO categorizes health and other related factors based on their functionality (e.g. body function, ability to perform tasks and participation in daily life activities). One can frequently find multiple health conditions or comorbidities in the elderly, which, in turn, lead to a set of concomitant deficiencies able to influence the disability process. The failure in executing a given task results from a mismatch between the individual capacity and the demand of the task itself, which, in turn, are influenced by the environment and how the task is executed. A common example is the ability to perform a brisk walking, which can be an impossible task to do in middle air, but that can still be achievable within an aquatic environment or may benefit from auxiliary tools, such as a walker. This allows to understand that the ability to perform a particular activity is related to many factors and depends upon the capacity of the elderly, which can be somehow assisted (e.g. walker), or by reducing the requirement of the task. This is a key aspect in the whole process of increasing/maintaining the functional capacity of the elderly (Castro 2015).

Physical activity fosters actual improvements in the elderly's capacity and body properties, namely the strength, flexibility, endurance and overall physical fitness. However, it is necessary to adapt the physical activity recommendations for older people to meet their particular needs, in the same way that it is essential to use specific types of exercises. Physical therapists can contribute to the design of physical activity programmes by modifying the exercises or using a compensation mechanism, which might provide an effective gain and response to the individual need of each old person. In the previous example, the brisk walking, one can easily understand that, whenever an elderly is incapable of performing the task autonomously, he/she will not do it (Castro 2015). However, the use of a walker and/or changes made in the task may allow the elderly to carry out the exercise, thus allowing to improve his/her strength, endurance and balance, among other components.

It is in this perspective that one must focus in achieving an outcome that, even though modest, will be the difference between whether or not the exercises can be accomplished in an autonomous manner. Often, it is difficult to understand the real meaning of measuring the physical ability. It could be said that a gain of over 5 metres in a distance travelled by an elderly over the same period of time does not appear to be a significant gain. However, if with this increase the elderly is able to get out of the room and move to other rooms of the house, then this gain also represents a huge gain in terms of functional capacity. On the other hand, this gain in the functional capacity may also mean an increased chance to participate in social tasks. For this reason, the assessment should be properly contextualized to better reflect the reality (Castro 2015).

1.4 Disorders and Psychomotor Rehabilitation

The general recommendations of physical activity programmes for the elderly may need to suffer some modifications, so they can efficiently contribute to reduce functional limitations. Some types of exercise may be particularly beneficial for certain individuals, while others may not. For frail elderly, a programme that includes resistance training and strength training proves to produce significant improvements in the overall functional capacity (Cuoco et al. 2004). Understandably, elderly with disabling medical conditions may be unable to fulfil the minimum recommended amount of physical activity, or may even be unable to adequately perform the exercises; however, this should not be a reason to not fulfil all activities within their limitations. Reducing sedentary behaviours by maintaining a certain routine that cuts down the number of resting hours can be a very reasonable plan for elderly that might be unable to perform other exercises (Sparling et al. 2015). The desired outcomes for a given activity programme will inevitably need to evolve with the changing characteristics of individuals. It is not infrequent that the clinical conditions change, negatively, in the older age and this should not be considered as an absolute impossibility to carry out some activities, but rather consider adapting them. For instance, cardiovascular conditions have a significant prevalence with strokes, but also acute cardiac conditions, such as ischemic heart disease and acute myocardial infarction. However, other chronic medical conditions, such as lung disease and osteoporosis, or metabolic changes, have major impact in the worldwide population, as we will see in the next sections (Castro 2015).

1.4.1 Cerebrovascular Accidents

In the case of strokes, the central nervous system control changes, predominantly presenting reflexive motor responses. Furthermore, muscle weakness and the degree of pathological changes in the muscle tone is expected. This whole process, as it affects one half of the body, hinders the normal movement and affects the most basic motor functions, such as walking. Additionally, in many cases, physical rehabilitation therapy in hospital settings is interrupted after 6 months and, unfortunately, often discontinued for most individuals. However, the potential to recover, although smaller, still remains and should be explored by maintaining the physical activity programmes. The practice of moderate to vigorous daily physical exercises for 40 min is recommended in individuals who have suffered from an ischemic stroke and that are still capable of performing regular exercise (Kernan et al. 2014). The general principles to conduct the exercises are based on the muscle tone maintenance during normal movement, exploring the performed exercise symmetrically, with loads, and progressively introducing complex movements, namely involving opposing movement patterns, such as alternatively stretching and bending the legs.

It is common to find movement patterns preventing these individuals from performing normal movements. A good example of such movement pattern is the full flexion of the upper limb that prevents the individual to grasp and drop an object, since the elderly is unable to simultaneously stretch the wrist and bend the fingers (Castro 2015). As such, the position in which the exercises are carried out will depend, inevitably, from the activity level of the individual, but the same principle will be kept for all positions. If we take the example of the laying position, the bridge exercise can be an excellent starting point to progress towards the walking movement. This position allows to simultaneously exercise both lower and upper limbs, while training different movement patterns, wherein the hip and the knee will stretch, while the ankle remains bended. While the individual improves, it becomes possible to increase the amount of load on the affected side and also the balance required to reduce the load on the opposite side. When the weight transfer becomes possible, the exercise can be used as a starting point for the transition to a sitting position. A key aspect in this and other similar positions is the need to actively strengthen the trunk muscles, exercising the deep muscles, such as the transverse abdomen, internal and external obliquus, the perineal muscles, the glutes and the iliac psoas. One of the most frequent mistakes in this exercise is the accentuation of the lumbar lordosis that hides an ineffective performance and moves away from the initial objective of improving normal moving charge and opposite movement patterns in different joints. Although it might seem an extremely modest exercise, in fact, it is a powerful tool to prepare for the standing position and to start walking (Castro 2015).

Another relevant exercise for this situation is to transfer the weight and the equilibrium under different positions. Consider the sitting position, in which the weight transfer can be performed for the right and left sides, or forward and backward, and also on any combination of these quadrants. The weight transfer to one side induces an increase of the muscle activity on that side so as to support the load and a slight rotation of the trunk to the opposite side. Walking produces a similar effect in our body and, therefore, using a relatively stable position (e.g. sitting), these movement patterns will contribute towards it. The weight transfer in a sitting position can be carried out during several years, ranging from moving objects from one side to another, until the use of unstable surfaces, such as Bobath balls or rocking chair. It should be noted, again, that the progress of each exercise should be taken into account, starting from the simplest to the most complex, from fewer simultaneous moves to larger, from lighter to heavy loads and so on. Of course, the next step is to move to a standing position, starting from a sitting position. The necessary training to shift from sitting to standing positions in individuals suffering from hemiparesis, namely occurring from a stroke, reveals itself as a key motor control exercise, requiring the simultaneous coordination between antagonistic movements. In the early stages, we have a transfer of weight from the buttocks to the feet, keeping a triple flexion pattern in the legs, i.e. dorsiflexion of the ankle, knee flexion and hip flexion. Then, while standing and with the weight on the feet, it is necessary to stretch the knee and hip without extending the ankle, strengthening the lower limb muscles (Castro 2015).

1.4.2 Heart Disease

A greater emphasis on joint and organic mobilization, as well as an increased duration on the training, should be considered to improve the physical capacity of elderly suffering from coronary heart disease. This is necessary to provide a more effective response of the musculoskeletal system and cardio-respiratory readiness (Jonsson et al. 1990). Similarly, the "cooling" of the body is of major importance in this group of individuals, so as to provide a gradual dissipation of the thermal load and gradual reduction of the peripheral vasodilation induced by exercise. These individuals are at an increased risk of hypotension, caused by vasodilation due to the delay in the response of baroreceptors due to ageing. For this population, the heart rate takes longer to the rest values and, therefore, a greater rest period between the various components of the exercise, or alternating between low and high intensity activities, become necessary. Aerobic exercise starts with low intensity, gradually increasing both intensity and duration. The absence of angina does not preclude the possibility of exercise-induced ischemia, and dyspnoea is often considered as an equivalent parameter. In these cases, the exercise intensity should correspond to 60–75% of maximum heart rate obtained in the stress test. This intensity is still effective to improve the aerobic metabolism and strength, while still providing greater comfort and less skeletal muscle difficulties (Pollock et al. 1991).

Lower intensity exercise for a longer period of time is safer and has better levels of adherence in long-term programmes. Still, the literature fails on presenting an ideal physical activity programme for coronary patients, only confirming that the exercises should be dynamic, enjoyable and accessible, without causing adverse consequences (Castro 2015). The classic example is a regular walk with the progressive increase in pace and distance travelled, to which upped body exercises must be added. The strength training, designed to improve muscle function and increase muscle mass, also increases the aerobic capacity and is an additional component in the rehabilitation of elderly suffering from coronary disease (Fiatarone et al. 1990; Kenner 1991). Hence, exercise intensity can be controlled by measuring the pulse, by the subjective perception of exertion, or by the stress test described above (Castro 2015).

1.4.3 Pulmonary Disease

It is common to observe a decrease in the physical activity of individuals with lung disease because of dyspnoea, but this can easily become a vicious cycle, as dyspnoea inversely increases with activity. Pulmonary rehabilitation aims to break this cycle by decreasing dyspnoea, thus promoting a better quality of life (Ries et al. 2007; Puhan et al. 2009; Bianchi et al. 2011; McCarthy et al. 2015). Resistance training is primordial for elderly suffering from this pathology and may consider both lower and upper limbs (Ringbaek et al. 2008). Even though the use of lower limbs has been more widely studied, a major part of daily activities, causing dyspnoea, are made using the upper limbs muscles, which also accumulate a respiratory function (Celli 1994; Ries et al. 1988).

For upper limbs exercising, the use of weights can bring additional benefits (O'Hara et al. 1984), although it is not clear which exercise improves endurance. Whenever possible, high-intensity continuous exercises should be adopted. However, as this may not be feasible for most of the elderly population, rest periods must be introduced (Coppoolse et al. 1999; Puhan et al. 2006; Beauchamp et al. 2010; Zainuldin et al. 2011). Given the rapid and shallow breathing pattern that these individuals present, it is mandatory to train the breath while performing the exercise so that it is slower and deeper (Holland et al. 2012), while simultaneously working the muscles involved in breathing and promoting its elongation (Castro 2015).

1.4.4 Osteoporosis

High-impact exercises carried out for 30 min, five to seven times per week, have shown beneficial effects on the bone density. This can be achieved with several types of exercises, including resistance training, running and jumping. Regarding the intensity, scientific evidences do not show clear advantages in the use of higher intensity exercise, such as running, compared to lower intensity, such as walking, and it is, therefore, preferable to adopt the type of exercise that better suits the elderly so it can be carried out regularly and systematically (Castro 2015).

1.4.5 Metabolic Disorders (Diabetes)

Exercise is recommended in type-1 and type-2 diabetes, with the objective of improving glycemic control, maintain the weight and reduce the risk of cardiovascular disease and overall mortality (Buse et al. 2007). This population must perform 30–60 min of moderate aerobic activity, avoiding more than two consecutive days of inactivity. If there are no contraindications, such as mild or severe proliferative retinopathy, strength training with loads should be carried out at least twice a week for the major muscle groups of the trunk, upper and lower limbs, carried out on 10 series repetitions (Gibala et al. 2012). Caution should be advised whenever performing intense isometric exercises that can cause a significant increase in blood pressure and precipitate intraocular haemorrhage. One should also avoid any kind of vigorous exercises in subjects with recent or active retina bleeding, and traumatic loads, such as long-distance running, or long downhill skiing, in diabetics with neuropathy, given the risk of stress fractures in the foot and ankle bones and the development of pressure sores on the fingers and toes (Castro 2015). It is also important to maintain relatively high fluid intakes, as well as measurement of glucose levels in the blood before, during and after exercise (Colberg et al. 2010).

1.5 Conclusions and Practical Implications

New challenges have been rising as the ageing population grows. In this sense, we should "listen to the elders" in the definition of policies to their social integration, as well as the development of initiatives aimed at increasing the quality of life and health of this population. It is true that the effects of ageing will become more pronounced all over the world in the coming years. Faced with this harsh reality, poverty and low income in this segment of the population will have repercussions in all countries and is necessary to define effective strategies to seek the welfare of elders. Another priority should be given on providing significant roles to the elderly, in order to foster their integration within their social context. This strategy should go through their integration in social, economic, cultural, sports and civil matters in our society. These aspects fall under the active ageing perspective, since they enable a comprehensive approach that includes family life, employment, education, integration, health and quality of life.

Physical activity has great benefits for the elderly, with the potential to reduce the causes of morbidity and increase the longevity of individuals. Whatever the age, it is never too late to become physically active and still get health benefits. Even exercises adapted to meet the functional limitations of the elderly, with limited intensity and slow progression, become very useful in dependent individuals or those with multiple comorbidities. In spite of this, exercise appears to be extremely important in motor rehabilitation process of older individuals with various diseases and may contribute, unequivocally, to maintain the health benefits. To this end, it is necessary to develop an individualized plan, with specific instructions, regarding the physical activity to be carried, the care, the precautions and the recommendations on activity progression, incorporating both preventive and therapeutic activities, in order to bring more life to years.

References

American College of Sports Medicine (ACSM). (2010). *ACSM's guidelines for exercise testing and prescription*. Philadelphia Wolters Kluwer/Lippincott Williams Wilkins.

Beauchamp, M. K., et al. (2010). Interval versus continuous training in individuals with chronic obstructive pulmonary disease—a systematic review. *Thorax, 65*, 157–164.

Bianchi, R., et al. (2011). Impact of a rehabilitation program on dyspnea intensity and quality in patients with chronic obstructive pulmonary disease. *Respiration, 81*, 186–195.

Blumenthal, J. A., et al. (1989). Cardiovascular and behavioral effects of aerobic exercise training in healthy older men and women. *Journal of Gerontology, 44*, 147–157.

Borg, G. (1982). Psychophysical bases of perceived exertion. *Medicine and Science in Sports and Exercise, 14*, 377–381.

Bowling, A. (2007). Aspiration for older age in the 21st century: What is successful ageing? *International Journal of Ageing and Human Development, 64*(3), 263–297.

Buse, J. B., et al. (2007). Primary prevention of cardiovascular diseases in people with diabetes mellitus: A scientific statement from the American Heart Association and the American Diabetes Association. *Circulation, 115*, 114–126.

Cabral, M. V., & Silva, P. A. (2009). *O Estado da Saúde em Portugal*. Imprensa de Ciência Sociais: Lisboa.

Cabral, M. V., & Silva, P. A. (2010). *A Adesão à Terapêutica em Portugal—Atitudes e Comportamentos da População Portuguesa Perante a Prescrição Médica*. Imprensa de Ciências Sociais: Lisboa.

Cabral, M. V., Silva, P. A., & Mendes, H. (2002). *Saúde e Doença em Portugal—Inquérito aos Comportamentos e Atitudes da População Portuguesa Perante o Sistema Nacional de Saúde*. Imprensa de Ciências Sociais: Lisboa.

Cabral, M. V., Ferreira, P. M., Silva, P. A., Jerónimo, P., & Marques, T. (2013). *Processos de envelhecimento em Portugal, Usos do tempo, redes sociais e condições de vida*. Fundação Francisco Manuel dos Santos: Lisboa.

Capucha, L. (2005). Envelhecimento e políticas sociais: Novos desafios aos sistemas de protecção. Protecção contra o "risco de velhice: Que risco?". Porto: Faculdade de Letras da Universidade do Porto.

Castro, M .A. (2015). Importância da actividade física no processo de reabilitação. In G. Dias, R. Mendes, P. Serra e Silva, & M. A. Banquinho (Eds.), *Envelhecimento activo e actividade física* (pp. 101–119). Coimbra: Escola Superior de Educação de Coimbra.

Celli, B. (1994). The clinical use of upper extremity exercise. *Clinical Chest Medicine, 15*, 339–349.

Colberg, S. R., et al. (2010). Exercise and type 2 diabetes: The American College of Sports Medicine and the American Diabetes Association: Joint position statement. *Diabetes Care, 33*, 147–167.

References

Comissão Europeia. (2012). *Concretizar o Plano de Execução Estratégica da Parceria Europeia de Inovação para um Envelhecimento Ativo e Saudável*. Brussels: European Commission.

Coppoolse, R., et al. (1999). Interval versus continuous training in patients with severe COPD: A randomized clinical trial. *European Respiratory Society, 14*, 258–263.

Cruz, P. (2009). *Envelhecimento Activo: Mudar o presente para ganhar o futuro*. REAPN—Rede Europeia Anti-Pobreza.

Cuoco, A., et al. (2004). Impact of muscle power and force on gait speed in disabled older men and women. *Journal of Gerontology Series A: Biological Sciences and Medical Sciences, 59*, 1200–1206.

Dias, G., & Mendes, R. (2013). Atividade Física para a Terceira idade. Coimbra.

Dias, G., Mendes, R., Serra e Silva, P., & Banquinho, M.A. (2014). Envelhecimento Activo e Actividade Física. In G. Dias, R. Mendes, P. Serra e Silva, M. A. Banquinho (Eds.). Coimbra: Escola Superior de Educação de Coimbra.

Dias, G., Mendes, R., Serra e Silva, P., & Banquinho, M.A. (2015). Gerontomotricidade: Actividades lúdicas e pedagógicas para o corpo envelhecido. In G. Dias, R. Mendes, P. Serra e Silva, M. A. Banquinho (Eds.). Coimbra: Escola Superior de Educação de Coimbra.

Fernández-Ballesteros, R. (2002). Social Support and Quality of Life Among Older People in Spain. *Journal of Social Issues, 58*(4), 645–659.

Fernández-Ballesteros, R., Garcia, L. F., Abarca, D., Blanc, E., Efklides, A., Moraitou, D., et al. (2010). The concept of 'ageing well' in ten Latin American and European countries. *Ageing & Society, 30*, 41–56.

Fiatarone, M. A., et al. (1990). High-intensity strength training in nonagenarians. Effects on skeletal muscle. *Jama, 263*, 3029–3034.

Fleg, J. L., & Lakatta, E. G. (1988). Role of muscle loss in the age-associated reduction in VO2 max. *Journal of Applied Physiology, 65*, 1147–1151.

Fonseca, A. M. (2005). *Desenvolvimento Humano e Envelhecimento*. Climepsi Editores: Lisboa.

Gibala, M. J., et al. (2012). Physiological adaptations to low-volume, high-intensity interval training in health and disease. *Journal of Physiology, 590*, 1077–1084.

Holland, A. E., et al. (2012). Breathing exercises for chronic obstructive pulmonary disease. *Cochrane Database of Systematic Reviews, 10*, CD008250.

Jonsson, P. V., et al. (1990). Hypotensive responses to common daily activities in institutionalized elderly. A potential risk for recurrent falls. *Archives of Internal Medicine, 150*, 1518–1524.

Kenner, T. (1991). Exercise in Health and Disease: Evaluation and Prescription for Prevention and Rehabilitation. Philadelphia W. B. Saunders.

Kernan, W. N., et al. (2014). Guidelines for the prevention of stroke in patients with stroke and transient ischemic attack: A guideline for healthcare professionals from the American Heart Association/American Stroke Association. *Stroke, 45*, 2160–2236.

Keysor, J. J. (2003). Does late-life physical activity or exercise prevent or minimize disablement? A critical review of the scientific evidence. *American Journal of Preventive Medicine, 25*(3 Suppl 2), 129–136.

King, A. C., & Guralnik, J. M. (2010). Maximizing the potential of an aging population. *JAMA, 304*, 1944–1945.

Lee, F., Mackenzie, L., & James, C. (2008). Perceptions of older people living in the community about their fear of falling. *Disability and Rehabilitation, 30*(23), 1803–11.

Lima, A. M. (1999). A Gestão da Experiência de Envelhecer em um Programa para a Terceira Idade. *Centro de Referência e Documentação sobre Envelhecimento, da Universidade Aberta da Terceira Idade—UnATI, 2*(2), 22–63.

Lobo, A., et al. (2011). Comparison of functional fitness in elderlies with reference values by Rikli and Jones and after one-year of health intervention programs. *Journal of Sports Medicine and Physical Fitness, 51*, 111–120.

Magalhães, E. E. (2005). Envelhecimento Demográfico Novos Desafios, *Workshop—UTAD*. Instituto Politécnico de Bragança, Escola Superior de Saúde.

McCarthy, B., et al. (2015). Pulmonary rehabilitation for chronic obstructive pulmonary disease. *Cochrane Database of Systematic Reviews, 2*, CD003793.

McGuire, D. K., et al. (2001). A 30-year follow-up of the Dallas Bedrest and Training Study: I. Effect of age on the cardiovascular response to exercise. *Circulation, 104*, 1350–1357.

Nations, U. (2001). World Population Ageing: 1950–2050. *Department of Economic and Social Affairs: Population Division, Magnitude and Speed of Population Ageing*, 11–13.

Neda, A., Silversein, M., & Parker, M. G. (2011). Late-life and earlier participation in leisure activities: Their importance for survival among older persons. *Adaptation and Aging, 35*(3), 210–222.

Nelson, M. E., et al. (2007). Physical activity and public health in older adults: Recommendation from the American College of Sports Medicine and the American Heart Association. *Medicine and Science in Sports and Exercise, 39*, 1435–1445.

O'Hara, W., et al. (1984). Weight training and backpacking in chronic obstructive pulmonary disease. *Respiratory Care, 29*, 1202–1210.

Oliveira, B. (2008). *Psicologia do Envelhecimento e do Idoso*. LivPsic: Porto.

OMS. (2001). *Active ageing: A policy framework*. Genève.

OMS. (2002). Rumo a uma linguagem comum para Funcionalidade, Incapacidade e Saúde.

OMS. (2005). *Envelhecimento Activo: Uma política de saúde*.

ONU (1998). *Department of economic and social affairs: World population projections to 2150*.

ONU (2001). *Department of economic and social affairs: World population prospects: The 2000 Revision*.

Pahor, M., et al. (2006). Effects of a physical activity intervention on measures of physical performance: Results of the lifestyle interventions and independence for Elders Pilot (LIFE-P) study. *Journal of Gerontology Series A: Biological Sciences and Medical Sciences, 61*, 1157–1665.

Paúl, C. (2005). Envelhecimento activo e redes suporte social. *Revista Sociologia, 15*, 275–287.

Pollock, M. L., et al. (1991). Injuries and adherence to walk/jog and resistance training programs in the elderly. *Medicine and Science in Sports and Exercise, 23*, 1194–1200.

Province, M. A., et al. (1995). The effects of exercise on falls in elderly patients. A preplanned meta-analysis of the FICSIT Trials. Frailty and injuries: Cooperative Studies of Intervention Techniques. *JAMA, 273*, 1341–1347.

Puhan, M., et al. (2009). Pulmonary rehabilitation following exacerbations of chronic obstructive pulmonary disease. *Cochrane Database of Systematic Reviews, 1*, CD005305.

Puhan, M. A., et al. (2006). Interval versus continuous high-intensity exercise in chronic obstructive pulmonary disease: A randomized trial. *Annals of Internal Medicine, 145*, 816–25.

Quintela, M.J. Grupo de Intervenção comunitária (GIC) -Idosos da Fundação Portuguesa de Cardiologia. Howell, F., Riniotis, K., Beech, R., Crome, P., & Ong, B.N. (2009). Older people and falls: Health status, quality of life, lifestyle, care networks, prevention and views on service use following a recent fall. *Journal of Clinical Nursing, 18*(16), 2261–72.

Ries, A. L., et al. (1988). Upper extremity exercise training in chronic obstructive pulmonary disease. *Chest, 93*, 688–92.

Ries, A. L., et al. (2007). Pulmonary rehabilitation: Joint ACCP/AACVPR evidence-based clinical practice guidelines. *Chest, 131*(5 Suppl), 4S–42S.

Ringbaek, T., et al. (2008). Rehabilitation in COPD: The long-term effect of a supervised7-week program succeeded by a self-monitored walking program. *Chronic Respiratory Disease, 5*, 75–80.

Serra e Silva, P (2012). *Aprender a Não Ser Velho*. Coimbra.

Serra e Silva, P., Dias, G., Branquinho, M. A., & Mendes, R. (2015). Envelhecimento activo. In G. Dias, R. Mendes, P. Serra e Silva, & M. A. Banquinho (Eds.), *Envelhecimento activo e actividade física* (pp. 1–21). Coimbra: Escola Superior de Educação de Coimbra.

Shephard, R. (1989). Habitual physical activity levels and perception of exercise in the elderly. *Journal of Cardiopulmonary Rehabilitation, 9*, 17–23.

Sherrington, C., et al. (2004). Physical activity interventions to prevent falls among older people: Update of the evidence. *Journal of Science and Medicine in Sport, 7*, 43–51.

Sparling, P. B., et al. (2015). Recommendations for physical activity in older adults. *BMJ, 350*, h100.

References

Spirduso, W. W. (2005). *Physical dimensions of aging* (2nd ed.). Champaign: Human Kinetics.

Stella, A. (2008). O Envelhecimento em Portugal no século XXI: Contributos e Reflexões de Prospectiva Demográfica. *Revista Cadernos de Economia, 84*, 1–10.

Stessman, J., et al. (2009). Physical activity, function, and longevity among the very old. *Archives of Internal Medicine, 169*, 1476–83.

Takahashi, T., et al. (2002). Influence of cool-down exercise on autonomic control of heart rate during recovery from dynamic exercise. *Frontiers of Medical and Biological Engineering, 11*, 249–59.

Wolf, S. L., et al. (1996). Reducing frailty and falls in older persons: An investigation of Tai Chi and computerized balance training. Atlanta FICSIT Group. Frailty and Injuries: Cooperative studies of intervention techniques. *Journal of the American Geriatrics Society, 44*, 489–497.

Wolfson, L., et al. (1996). Balance and strength training in older adults: Intervention gains and Tai Chi maintenance. *Journal of the American Geriatrics Society, 44*, 498–506.

Zainuldin, R., et al. (2011). Optimal intensity and type of leg exercise training for people with chronic obstructive pulmonary disease. *Cochrane Database of Systematic Reviews, 11*, CD008008.

Zijlstra, G. A., van Haastregt, J. C., van Rossum, E., van Eijk, J. T., Yardley, L., & Kempen, G. I. (2007). Interventions to reduce fear of falling in community-living older people: A systematic review. *Journal of the American Geriatrics Society, 55*(4), 603–15.

Chapter 2
Physical Activity Benefits in Active Ageing

Gonçalo Nuno Figueiredo Dias, Micael Santos Couceiro,
Pedro Mendes and Maria de Lurdes Almeida

Abstract Ageing is a biosocial process which results in the loss of capacity over time. This decline is gradual and can significantly vary from individual to individual, according to their genetic, morphological and functional characteristics. This chapter shows that regular physical activity can reduce the risk of prevalence of chronic diseases in agieng and reduce the risk of morbidity and mortality of the elderly. For example, physically active elderly can have higher levels of functionality, improved cognitive performance and a lower risk of falling. On this basis, a physical activity programme can minimize the motor decline in the elderly, preventing the loss of functionality or inability of states, thus promoting significant health benefits. One can also state that more physically active older individuals have lower ratios of mortality, heart disease, high blood pressure, stroke, type-2 diabetes, colon cancer and breast cancer.

Keywords Physical activity · Active ageing · Health · Quality of life

2.1 Background

The scientific community point out that regular physical activity can reduce the risk of prevalence of chronic diseases with ageing and, likewise, mitigate the physiological changes imposed at this stage. Furthermore, it additionally reduces the morbidity and mortality risks coming with ageing (Spirduso 2005; Paterson et al. 2007). On the other hand, regular physical activity enhances the optimization of cardio-respiratory and locomotor (e.g. skeletal muscle) systems, ensuring a healthier body composition (Bauman et al. 2005; Paterson and Warburton 2010). Accordingly, the physically active elderly can have higher levels of functionality, improved cognitive performance and a lower risk of falling, typically associated with balance issues. On this basis, a physical activity programme can minimize the motor deterioration of the elderly, preventing the loss of functionality or inability, thus promoting significant health benefits. Therefore, there is a strong scientific evidence that physically active older individuals have lower ratios of mortality, heart disease, high blood pressure, stroke, type-2 diabetes, colon cancer and breast cancer (Steiner et al. 2004; Mendes et al. 2015).

Several authors suggest that there is a positive association between cardio-respiratory fitness and functional capacity of the elderly (e.g. Paterson and Warburton 2010; Singh 2002; Dias et al. 2015). However, only some few studies have explored the minimum threshold and the optimum physical activity level for this old population (see Singh 2002). Still, these works consider that a moderately vigorous intensity is at the essence to effectively preserve and/or enhance health benefits. To this end, the prescription of exercise should favour activities that foster improvements in the functional capacity and independence of the elderly, slowing down morbidity and mortality (Kemper 2006). Adducing further contributions to this theme, various entities and international organizations, such as ACSM (1998); Nelson et al. (2007) and WHO (2010), present a set of recommendations for the elderly, advising moderately vigorous activities (e.g. brisk walking), maintaining the muscle mass through strength and power exercises, stimulating specific muscle groups, practicing balance and flexibility, etc. Given the above, we hereafter present the bio physiological and cognitive changes occurring during ageing, showing the extent to which physical activity can delay the deleterious effects associated with this phenomenon (Mendes et al. 2015).

2.2 Morphological and Functional Changes of the Elderly

Ageing is a biosocial process which results in the loss of capacity over time. This decline is gradual and can significantly vary from individual to individual, according to their genetic, morphological and functional characteristics (Mendes et al. 2015). Ageing defines the decrease in the efficiency of body functions with a series of morphological, psychological, biochemical and functional changes occurring as a result of time converging towards a difficulty in responding to internal and external stimuli (Alba 1986). As an objective and subjective process, ageing can be defined, not only by demographic and biological aspects, but also due to the interaction between multiple factors, including social (Staab and Hodges 1997). These are translated into a succession of irreversible changes, such as the loss or decrease of the social, family and labour roles, as well as the limitation of social relationships (Berger 1995b). It should be noted that Rdwanski and Hoeman (2000) consider that the fact of living longer is a relevant paradox between retirement and enjoying the golden years with an 80% probability of getting one or more chronic disabling diseases. Consequently, chronic diseases and physiological age changes increase the likelihood of physical limitations and disabilities of older people in a disproportional manner when compared to young adults. As a result, ageing is not only related with the decline of biological-organic operation, but also with the ability to adapt to the environment be as functional as possible (Hoeman 2000; Mendes et al. 2015).

2.3 Body Composition

Besides the visible changes in the physical appearance, changes in the body composition of the elderly have implications in their physical function and health. In this context, fat redistribution and muscle mass loss result in a significant reduction of the aerobic capacity. The body weight begins to stabilize at the age of 50 and tends to decrease during the seventh decade of life, but the body fat still rises and the basic gynaecoid and android features are kept (Spirduso 2005). The fat is somehow redistributed with ageing, however, differently for the two genres. For example, in men, subcutaneous fat decreases on the periphery of the body, but the deposit increases both centrally and internally. In women, total fat increases, especially the internal body fat (visceral) (Mendes et al. 2015).

2.4 Cardio-respiratory Capacity

With ageing, inspiratory capacity decreases as a result of intercostal cartilage calcification, with a reduction of contractility of the inspiratory muscles, loss of elasticity of the lung tissue, and weakening of the diaphragm and intercostal muscles. Also, the residual capacity (volume of air that remains in the respiratory system at the end of a normal exhalation) increases, while the vital capacity (maximum amount of air a person can expel from the lungs after a maximum inhalation) decreases by about 25%. The total lung capacity (the sum of both vital and residual capacities), measured during a forced inspiration, decreases to 50% between the ages of 20 and 80 (Berger 1995a; Spirduso 2005; Mendes et al. 2015). As the person gets older, other physiological changes, particularly musculoskeletal, lead to a gradual decrease in lung function. Associated with a reduced respiratory muscle strength, the inner elastic fibres located within the alveolar walls (terminal bronchioles) lose their elasticity. This phenomenon gives rise to a deterioration of the airways, affecting the alveolar ventilation (Reis 1995; Vicente 2012).

Other biological changes, such as decreased cardiac output, increased blood pressure, reduction of the maximum heart rate, and increased peripheral vascular resistance, are also related with the decline in oxygen consumption. Concomitantly, other variables can contribute to a deficit in the ability to produce muscular activity, including the reduction of amplitude and frequency of respiratory movements and the exhaled air volume (Correia and Silva 1999). Furthermore, cardiovascular fitness inevitably decreases with age, and the changes in body composition and cardiorespiratory system, together with decreased physical activity during ageing, are responsible for most of the cardiovascular fitness decline (Mendes et al. 2015).

2.5 Musculoskeletal System

The literature shows that muscle parameters, such as the number of fibers, the size and diameter, degree and contraction speed, decrease inversely proportional to the use made by each segments (Singh 2002; Spirduso 2005). As such, changes in muscle strength, over the years, seem to be dependent on the muscle itself (Correia and Silva 1999; Carvalho and Soares 2004). The reduction of strength in elderly patients is due to essentially two factors: (1) Loss of muscle mass because of atrophy and reduction in the number of fibres; and (2) Metabolic changes in contractile proteins. It is noteworthy that the loss of strength in the lower limbs of the elderly is particularly evident, especially on the proximal muscles of the hip and thigh since they are particularly affected by the muscle fibre atrophy and the maximum strength decline (Correia and Silva 1999). This reduced muscle strength in the lower limbs is limiting in everyday tasks, affecting gait efficiency, climbing stairs, or get out of bed or a chair, having been identified as one of the factors associated with increased risk falls (Mendes et al. 2015).

More important than simply the loss of the elderly's maximum force, for Correia and Silva (1999), in a functional point-of-view, is the loss of muscle power, since fast strength is the manifestation of the strength necessary to carry out daily life activities, such as walking, climbing stairs and lifting objects. The loss of muscle power in the elderly manifests itself in the time needed to reach the maximum peak force, and the contraction and relaxation times. The reduction of muscle mass has been suggested as the primary cause for the loss of muscle strength with age, as well as power, speed, flexibility and precision of movement (Mendes et al. 2015). However, the loss of functional ability is not only due to the degeneracy inherent to ageing, but it is also a result of the reduced use of muscles and consequent reduction in the stimulation (Freitas et al. 2002; Mendes et al. 2015).

2.6 Central Nervous System

The central nervous system is strongly affected by ageing, in particular at the level of the cell function and reduction of neurons, which leads to a reduction of fibres and nerve bundles, and a decrease in the transmission and/or reception capacity of the brain nervous influx (Mendes et al. 2015). This process starts very early (around the age of 50) and affects all nervous structures, including the brain and spinal cord. In this sense, the nerve modifications described in the literature are the following: (*i*) Brain Atrophy (loss of weight and volume reduction); (*ii*) Increase in connective tissue (senile plaques); (*iii*) reduction of blood supply and oxygen consumption in the brain; and (*iv*) Gradual increase of cerebral vascular resistance (Reis 1995). It should also be noted that the decrease in the number of neurons (neuronal death) causes a reduction of fibres and nerve bundles and a low transmission capacity or reception of nerve influxes to the brain. The reaction time increases and the response to stimuli takes place more slowly because of the change in the proprioceptors and decrease in the number of synapses (Fontaine 2000; Mendes et al. 2015).

2.7 Sensory and Perceptive System

The auditory perception of the elderly is often damaged, either on the auditory localization capabilities, or regarding its discrimination. In the elderly, the main hearing problems are related to physiological changes, such as: (*i*) degeneration of the hearing nerve fibre (cochlea and the organ of Corti); (*ii*) eardrum thickening; (*iii*) reduction in the production of cerumen; (*iv*) increasing in the stiffness of the middle ossicles (stapes); and (*v*) Atrophy of the auditory nerve (Berger 1995a; Roach 2003). All these conditions lead to hearing loss, which becomes more evident at the age of 50 and gets even worse after the age of 70. Presbycusis is the most common problem and is reflected by a decreased perception of pure tones. It begins with the inability to hear certain sounds, especially at high frequencies (treble). Then, as we get older, it gets more difficult to distinguish human voice from the environmental noise. The language becomes often unintelligible, and the normal tone of voice and the consonants are less noticeable than the vowels because they are located at a higher frequency (Berger 1995a; Vincente 2012; Mendes et al. 2015).

Like hearing, vision is a particularly sensitive sensory modality affected by ageing. The consequences of ageing in the visual system emerge progressively. For Berger (1995a), Vincente (2012) and Mendes et al. (2015), a first change is observed in the transmissivity of the eye and its accommodation capacity. This leads to problems in the perception of depth, sensitivity to glare and colours. Fontaine (2000) goes a step further and points out some details about visual perception changes, such as: (*i*) the perception of colours seems affected, especially in discriminating certain colours (the eye captures bright colours more easily, like yellow, red and orange, and finds it more difficult to capture blue, violet and green); (*ii*) visual acuity, which refers to the ability of resolution and discrimination, is effected (loss of fine detail vision); (*iii*) Adaptation, which is the sensitivity of the eye to changing light intensities (adaptation to darkness and light), decreases with age and is reflected by the increased adaptation time; (*iv*) The visual field, i.e. the area covered with visual fixation, which also decreases with age; and (*v*) the peripheral field decreases of some few degrees between after the age of 40–45.

The proprioceptive information system is yet another important change occurring with ageing. Eliopoulos (2011) shows that the elderly finds it harder to completely discriminate the limbs' movements; either active or passive. This lack of ability to properly recognize the position of the segments, allied with a reduced sensitivity and pressure, can cause issues in the postural control, especially under low light conditions. According to the same author, body posture, rather than being the result of a natural, simple regulation, becomes more of a problem, since the elderly's motion control requires an additional processing information and, of course, a longer treatment. The time needed to make necessary corrections over unforeseen constraints and variations is also longer with ageing. One way to overcome this is to walk slowly; a strategy that one may observe in the elderly population and one they claim to be of great importance to avoid accidents, especially falls (Almeida 2011; Mendes et al. 2015).

In sum, the senescence essentially covers three modalities: balance, hearing and sight. Ageing has important consequences, sometimes severe, at both psychological and social levels. Sensory deficits of auditory and visual nature seem to be important causes of the general decline in intellectual functioning. Some sensory modalities, like the sense of smell, is only slightly affected with age, while others, such as hearing and vision, are badly affected. Finally, slow movements are a simple reflection of the sensory decline: it is, actually, a natural consequence of ageing (Eliopoulos 2011; Vicente 2012; Mendes et al. 2015).

2.8 Movement Duration and Motor Reaction

Authors, such as Correia and Silva (1999), argue that the duration of a movement and the reaction time increase with the difficulty of the task and its novelty. These are due to the elderly's slower information processing, in which the duration increases based on different stages, namely: (*i*) treatment of sensory information; (*ii*) decision-making; and (*iii*) Movement programming. Although the elderly retains part of the psychomotor skills over the years, the motor system suffers a general delay; something that affects the reaction time over various daily life stimuli (Berger 1995b). The reaction time starts increasing around the age of 45–50 and stabilizes around the age of 70. The same author (Berger 1995b) also points to several factors that can influence the reaction time in older people, namely: (*i*) decreased eyesight and hearing; (*ii*) motor response delay to a sensory stimulation; (*iii*) memory loss; (*iv*) low motivation; (*v*) significant absence of social contact; and (*vi*) disease. In addition, reflexes get slower with ageing. This decrease in the effectiveness of reflexes, associated with the difficulty of the nervous system to process information, seems to explain the high rate of accidents in the elderly population, as they usually take longer to evaluate the surroundings and make effective decisions (Berger 1995b; Mendes et al. 2015).

Ageing also causes a significant decrease in the motor performance of the elderly, wherein the speed is a critical issue. Undeniably, in motor sequences that highly depend on the perceptual information, especially of external nature, the processing of such information shows significant delays. Likewise, in complex motor sequences, under significant time constraints, the neuronal and motor delays are notorious. Finally, when performing slow actions, which have been extensively practiced throughout life, one can observe a reasonable level of performance (Correia and Silva 1999).

Besides those changes occurring while ageing, the ability to receive information, the central process of preparing data and the peripheral systems to produce a given response, the motor organization strategies and the energy resources management, also change. Consequently, actions, movements and reactions are generally slower, inducing changes even in the movement patterns (Berger 1995b; Correia and Silva 1999; Godinho et al. 1999; Vieira and Koenig 2002; Yassuda 2002; Mendes et al. 2015).

2.9 Health Benefits of Physical Activity

The relationship between physical activity, health, quality of life and ageing has aroused interest of the scientific community. This is evident in the significant number of relevant works published in the past decades on the subject. As stated by Matsudo et al. (2001) and Mendes et al. (2015), there is a consensus among health professionals that physical activity is a major factor for the successfully ageing process. Some aspects about the prescription of exercises to the elderly are now highlighted.

2.9.1 Aerobic Capacity

The scientific community generally agrees that adults should carry out physical activity every week, preferably on a daily basis, with a minimum of 30 min of moderate intensity (Pate et al. 1995; Nelson et al. 2007). This comprises the elderly population, even more so as the inactivity tends to increase with age and may even reach 62% of the population over 65-years old (NACA 2006). The Native Americans for Community Action (NACA) even considers the inactivity as a highly relevant risk factor (NACA 2006). Many authors even believe in the relationship between regular physical activity and the rate of mortality and morbidity. As suggested by numerous epidemiological studies conducted in recent decades, the risk of morbidity and mortality decreases as the physical condition (cardio-respiratory) of the elderly increases (e.g. Blair et al. 1989, 2001; Lakka et al. 1994; Ekelund et al. 1988; Lie et al. 1985; Sandvik et al. 1993; Sobolski et al. 1987; Arraiz et al. 1992; Myers et al. 2004). Yet, finding the necessary amount of activity required to maintain the autonomy of the elderly still needs to be assessed. A thorough review work from Paterson et al. (2007) mentions the work of (Mendes et al. 2015) that, according to which, carrying out vigorous sports at least twice a week leads to a significant reduction in the risk of the coronary artery disease or mortality.

In spite of this, in the elderly case, the intensity of the aerobic load should range from moderate to vigorous, so as to reduce the death rate. While moderate intensity translates into a maximum oxygen consumption (VO_2max) between 40 and 60%, moderately strong intensity goes about 50 and 70% of V0$_2$max. The work developed by Martins et al. (2010) concluded that aerobic and other moderate to vigorous programmes enhance the metabolic health indicators (Mendes et al. 2015). Several other authors delimit the cumulative amount of energy expenditure in 4200 kJ/week (load volume), from 4 to 6 Metabolic Equivalent of Tasks (METs)[1] as necessary to delay mortality or reducing it by 25–30% (cf. Lee and Skerrett 2001). However, it is possible to establish a link between a slight increase in the physical activity, like increasing one additional hour a week of walking, and a reduced risk of mortality (Oguma et al. 2004). In this line of thought, consider that physical activity with low energy expenditure in older adults, aged between 70 and 82, significantly reduces the risk of mortality (32%). Also, Slentz et al. (2007) assessed the minimum level of physical activity required to maintain metabolic health, concluding that a volume of 13 km/week of physical activity, such as walking, would be enough for the elderly to keep the weight balanced (Matsudo et al. 2001).

We emphasize that the prescription of aerobic exercises so far presented is mainly aimed at the active and healthy elderly. However, for the frail or weakened elderly, the aerobic training load becomes considerably more difficult to manage because of the potential clinical constraints, such as gait pattern disorders, arthritis, dementia, cardiovascular diseases, orthopaedic problems, incontinence and changes in visual acuity. In these cases, the range of possible activities goes from the ergometer of upper limbs, lower limbs and even sitting exercises performed in water environment (Mendes et al. 2015).

2.9.2 Muscular Strength

Besides the aerobic work, the literature also emphasizes strength training as the key to improve the mobility of daily life activities and the overall autonomy of the elderly (Spirduso 2005). Several studies indicate an intensity of strength ranging between moderate to high, i.e. superior to 60%, with 1 maximum repetition (MR) for two days a week, to maximize the adaptive gains (Smilios et al. 2007; Wieser and Haber 2007; Paterson et al. 2007). Hunter et al. (1995) observed improvements in the functional abilities of female elderly (between 60 and 77 years) after 12 weeks of muscular training. According to Nelson et al. (2007), today's elderly can benefit from embodied training programmes in calisthenics (self-loading) and in the progressive method with resistance (e.g. elastic and weight machines). Regard resistive strength exercises, the same authors recommend, at least, a 10–15 repetitions series for large muscle groups, suggesting that, for each muscular group, the elderly should workout in two or three non-consecutive days per week (Mendes et al. 2015).

---

[1]MET is a unit used to quantify the intensity of physical activity.

Despite the recommendations for the elderly, issued by various authors and international organizations, to focus, preferentially, on aerobic activity programmes and muscular strength, it is the latter that will somehow stabilize or reverse the loss of muscle mass (e.g. sarcopenia). Therefore, this reinforces the priority role that muscular strength exercises play in maintaining the functional capacity and independence of the elderly (Matsudo et al. 2001). In groups of frail elderly and reduced ambulatory ability, Singh (2002) proposes a set of recommendations aimed at maintaining the motor functionality. The same author considers that the use of free weights, like dumbbells, wrist and knee weights, elastic bands, and the body's own weight (self-load), allows to develop a tough strength training with low to moderate intensities, but still valid for this type of elderly population. Although such training does not trigger significant physiological adaptations in the elderly, ultimately reducing disability (Morris et al. 1999), it does optimize the stability in gait pattern (Hausdorff et al. 2001), reduces the ratio of injuries caused by falls (Robertson et al. 2001), and decreases the pain resulting from arthritis and similar problems (Ettinger et al. 1997; Mendes et al. 2015).

2.9.3 Flexibility

As opposed to the motor skills already addressed, the specific benefits of flexibility for the elderly's health, do not "earn" the same consensus in the scientific community. However, the ACSM (1998) recommends exercises, such as walking, dancing and stretching that, when integrated in a physical programme, can contribute to an increased range of motion. As such, it is recommended to carry out, at least, 10 min of flexibility exercises, aimed at working the major muscle and tendon groups. For this, static methods can be considered, between 10 and 30s, in order to maintain the stretch in 3 or 4 series per exercise. This type of exercise should complement the strength and aerobic activities, but preceding it in order to avoid any muscle injuries (Franklin et al. 2000; Mendes et al. 2015).

2.9.4 Balance

One of the main causes of accidents and disabilities in the elderly are falls, which usually happen due to the lack of balance (Nelson et al. 2007; Paterson et al. 2007). According to Spirduso (2005), the risk of falling is represented by a multifactorial matrix and, hence, any intervention requires a multidisciplinary team (e.g. nurse, physical therapist, physiotherapist, pharmacist, occupational therapist, psychiatrist, psychologist and physical education professionals). This is not oblivious to the many factors that are normally associated with falls, especially: weakened muscles, falling historic, polypharmacy, deficit balance, the use of walking aids (e.g. walker), deficient vision, arthritis, deterioration of daily life activities, depression, deterioration of cognitive functionss and advanced age (over 80 years). Notwithstanding that, regular physical activity, by itself, can reduce falls and associated injuries in 35–45% (Robertson et al. 2002). As stated by the WHO (2010), there are good evidences that older adults with low mobility adhering to a regular and safe physical activity programme, can reduce the risk of falling by about 30%. Robertson et al. (2002) propose balance exercises three times a week.

2.9.5 Biopsychosocial Model

As many authors have stated, at the physiological level, exercise programmes improve the cardio-respiratory condition, as well as the strength and functionality of the elderly, slowing the motor decline. These programmes enhance the maintenance of both autonomy and quality of life. However, these benefits go beyond the biological and physiological aspects of this population (Matsudo 2001; Spirduso 2005). This stage of human development presents itself as a multifactor matrix, where the elderly emerges as a biopsychosocial being with needs and different capacities (Fonseca 1998; Spirduso 2005). In this respect, several factors are associated with the ageing process, such as molecular, cellular, systemic, behavioural, cognitive, social, among others (World Health Organization 1998; Palácios 2004; Mendes et al. 2015).

In the context of mental health, Paterson and Warburton (2010) build up the association between long-term regular physical activity with a decrease in the risk of dementia and Alzheimer's disease. For them, exercises can lead to an improvement in the cognitive function of the elderly. In spite of this, there are also evidences of the benefits of physical training on mood states (Engels et al. 1998; Martins et al. 2011). In the case of depression, this significantly affects the quality of life of the elderly and is, therefore, considered as a risk factor for dementia. As it is suggested by Singh (2002), regular physical activity potentiates positive psychological attributes in the elderly and can act as a non-pharmacological treatment of depressive disorders.

Also, physical exercises promote a greater commitment to a more active lifestyle, which is often associated with self-esteem and self-improvement (Stella et al. 2002). Socially, walking and organized physical sessions, appropriate to each elderly, leads to more social interaction, reducing loneliness and social exclusion. In this sense, the World Health Organization (1998) discloses the social benefits of physical activity highlighting, as immediate gains, the occupational engagement and consequent social and cultural integration, and, as long term, the possibility of new friendships, playing an active role, and improving the "inter-generational" activity (Mendes et al. 2015).

2.10 Conclusions and Practical Implications

The human body, over time, inevitably goes through a dynamic process of morphological and functional changes. In the old age, the reduction in the capacity of the organs and other functional systems can trigger pathological processes, reducing the quality of life of the elderly.

One of the problems contaminating any physical activity programme designed for the elderly is the fact that the functional gains earned during training are lost very quickly once the intervention ends. This suggests that every effort should be made to always intervene in the long term, thus increasing the motivation and retention of active senior citizens in such initiatives. We emphasize that active ageing requires a dynamic balance between body and mind. Thus, it is necessary to see the elderly as someone who thinks, feels and moves differently, something that requires special care in terms of body practices, as well as in the management of organic, nutritional and physiological aspects covering the ageing body.

References

Alba, S. A. (1986). Envejecimiento humano a nivel individual y de las poblaciones. In S. A. Alba, G. Llera, & J. D. Peña (Eds.), *Tratado de Geríatria y Asistencia Geriátrica* (pp. 15–28). Barcelona: Salvat Editores.

Almeida, M. L. (2011). Auto cuidado e promoção da saúde do idoso: contributo para uma intervenção em enfermagem. Tese de doutoramento em Enfermagem [On-line]. Consultado em 19 de maio de 2014 de http://repositorio-aberto.up.pt/handle/10216/69713.

American College of Sports Medicine Position Stand (ACSM). (1998). The recommended quantity and quality of exercise for developing and maintaining cardiorespiratory and muscular fitness, and flexibility in healthy adults. *Medicine and Science in Sports and Exercise, 30*, 975–91.

Arraiz, G. A., Wigle, D. T., & Mao, Y. (1992). Risk assessment of physical activity and physical fitness in the Canada Health Survey mortality follow-up study. *Journal of Clinical Epidemiology, 45*, 419–428.

Bauman, A., Lewicka, M., & Schöppe, S. (2005). *The health benefits of physical activity in developing countries*. Geneva: World Health Organization.

Berger, L. (1995a). Aspectos biológicos do envelhecimento. In L. Berger & D. Mailloux-Poirier (Eds.), *Pessoasidosas: Uma abordagem global* (pp. 123–156). Lisboa: Lusodidacta.

Berger, L. (1995b). Aspectos psicológicos e cognitivos do envelhecimento. In L. Berger, & D. Mailloux-Poirier (Eds.), *Pessoas idosas: Uma abordagem global* (pp. 157–197). Lisboa: Lusodidacta.

Blair, S. N., Cheng, Y., & Holder, J. S. (2001). Is physical activity or physical fitness more important in defining health benefits? *Medicine and Science in Sports and Exercise, 33*(6 Suppl.), S379–S399.

Blair, S. N., Kohl, H. W., Paffenbarger, R. S., Jr., Clark, D. G., Cooper, K. H., & Gibbons, L. W. (1989). Physical fitness and all cause mortality. A prospective study of healthy men and women. *JAMA, 262*, 2395–2401.

Carvalho, J., & Soares, J. M. (2004). *Envelhecimento e força muscular: breve revisão*. [On-line]. Obtido em 13 de Fevereiro de 2011, de Revista Portuguesa de Ciências do Desporto:http://www.projetosaudebh2011.com/news/envelhecimento%20e%20for%C3%A7a%20muscular%20-%20breve%20revis%C3%A3o/.

Correia, P. P., & Silva, P. A. (1999). Alterações da função neuromuscular no idoso. *Actas do Simpósio 99, Envelhecer melhor com a actividade física* (pp. 51–62). Edições da Faculdade de Motricidade Humana: Cruz Quebrada.

Dias, G., Mendes, R., Serra e Silva, P., & Banquinho, M.A. (2015). Gerontomotricidade: Actividades lúdicas e pedagógicas para o corpo envelhecido. In G. Dias, R. Mendes, P. Serra e Silva, M. A. Banquinho (Eds.). Coimbra: Escola Superior de Educação de Coimbra.

Ekelund, L. G., Haskell, W. L., Johnson, J. L., Whaley, F. S., Criqui, M. H., & Sheps, D. S. (1988). Physical fitness as a predictor of cardiovascular mortality in asymptomatic North American men. The lipid research clinics mortality follow-up study. *New England Journal of Medicine, 319*, 1379–1384.

Eliopoulos, C. (2011). *Enfermagem gerontológica* (7th ed.). Porto Alegre: Artmed.

Engels, H. J., Drouin, J., Zhu, W., & Kazmierski, J. F. (1998). Effects of low-impact, moderate-intensity exercise training with and without wrist weights on functional capacities and mood states in older adults. *Gerontology, 44*, 239–44.

Ettinger, W., Burns, R., Messler, S., et al. (1997). A randomized trial comparing aerobic exercise and resistance exercise with a health education program in older adults with knee osteoarthritis: The Fitness Arthritis and Seniors Trial (FAST). *Journal of the American Medical Association, 277*, 25–31.

Fonseca, V. (1998). *Psicomotricidade: Filogénese, ontogénese e retrogénese*. Diversos.

Fontaine, R. (2000). *Psicologia do Envelhecimento*. Lisboa: Climepsi Editores.

Franklin, B., Whaley, M., & Howley, E. (2000). *ACSM's Guidelines for Exercise Testing and Prescription*.

Freitas, E., Miranda, R., & Mônica, N. (2002). Parâmetros clínicos do envelhecimento e avaliação geriátrica global. In E. V. Freitas, L. Py, A. L. Neri, F. X. Cançado, M. Gorzoni, & S. Rocha (Eds.), *Tratado de geriatria e Gerontologia* (pp. 609–617). Rio de Janeiro: Guanabara Koogan.

Godinho, M., Melo, F., Mendes, R., & Chiviacowsky, S. (1999). Aprendizagem e envelhecimento. *Actas do Simpósio 99—Envelhecer melhor com a actividade física* (pp. 73–80). Serviço de Edições da Faculdade de Motricidade Humana: Cruz Quebrada.

Hausdorff, J., Nelson, M., Kaliton, D., et al. (2001). The etiology and plasticity of gait instability in older adults: A randomized controlled trial of exercise. *Journal of Applied Physiology, 90*, 2117–2129.

Hoeman, S. (2000). Lidar com doenças crónicas, incapacitantes ou do desenvolvimento. In S. Hoeman (Ed.), *Enfermagem de Reabilitação—Processo e aplicação* (2nd ed., pp. 209–250). Loures: Lusociência.

Hunter, G. R., Treuth, M. S., Weinsier, R. L., Kekes-Szabo, T., Kell, S. H., Roth, D. L., et al. (1995). The effects of strength conditioning on older women's ability to perform daily tasks. *Journal of the American Geriatrics Society, 43*, 756–760.

Kemper, H. C. G. (2006). Exercise and the physical consequences for the aging people. In J. Barreiros, M. Espanha, & P. P. Correia (Eds.), *Actividade Física e Envelhecimento* (pp. 121–134). Cruz Quebrada: Faculdade Motricidade Humana Serviço de Edições.

Lakka, T. A., Venalainen, J. M., Rauramaa, R., Salonen, R., Tuomilehto, J., & Salonen, J. T. (1994). Relation of leisure-timephysical activity and cardiorespiratory fitness to the risk of acute myocardial infarction. *New England Journal of Medicine, 330*, 1549–1554.

Lee, I. M., & Skerrett, P. J. (2001). Physical activity and all-causemortality: What is the dose-response relation?. *Medicine & Science in Sports & Exercise, 33* (Suppl.), S459–S471.

Lie, H., Mundal, R., & Erikssen, J. (1985). Coronary risk factors and incidence of coronary death in relation to physical fitness.Seven-year follow-up study of middle-aged and elderly men. *European Journal of Heart, 6*, 147–157.

Martins, R. A., Coelho e Silva, M., Pindus, D., Cumming, S., Teixeira, A., & Veríssimo, M. (2011). Effects of strength and aerobic-based training on functional fitness, mood and the relationship between fatness and mood in older adults. *Journal of Sports Medicine and Physical Fitness, 51*, 489–96.

References

Martins, R. A., Coelho e Silva, M. J., Teixeira, A. M., Veríssimo, M. T., & Cumming, S. P. (2010). Effects of aerobic and strength-based training on metabolic health indicators in older adults. *Lipids in Health and Disease, 9*, 76–81.

Matsudo, S. (2001). *Envelhecimento & Atividade Física*. Londrina: Midiograf.

Matsudo, S. M., Matsudo, V. K. R., & Neto, T. L. B. (2001). Atividade física e envelhecimento: aspectos epidemiológicos. *Revista Brasileira de Medicina Esporte, 7*, 2–13.

Mendes, P.C., M., Almeida, & Dias, G. (2015). Benefícios da actividade física no processo de envelhecimento individual. In: Envelhecimento Activo e Actividade Física (pp. 25–42). In G. Dias, R. Mendes, P. Serra e Silva, M. A. Banquinho (Eds.). Coimbra: Escola Superior de Educação de Coimbra.

Morris, J., Fiatarone, M., Kiely, D., et al. (1999). Nursing rehabilitation and exercise strategies in the nursing home. *Journal of Gerontology Medical Sciences, 54*A, M494–M500.

Myers, J., Kaykha, A., George, S., Abella, J., Zaheer, N., Lear, S., et al. (2004). Fitness versus physical activity patterns in predicting mortality in men. *American Journal of Medicine, 117*, 912–918.

NACA (2006). *Seniores in Canada 2006 report card* (Cat. N^{bre} HP30-1/2006E). National Advisory Council on Aging (NACA), Ottwa, Ontario.

Nelson, M. E., Rejeski, W. J., Blair, S. N., Duncan, P. W., Judge, J. O., King, A. C., et al. (2007). Physical activity and public health in older adults: Recommendation from the American College of Sports Medicine and the American Heart Association. *Medicine & Science in Sports & Exercise*, 1435–1445.

Oguma, Y., & Shinoda-Tagawa, T. (2004). Physical activity decreases cardiovascular disease risk in women: Review and meta-analysis. *American Journal of Preventive Medicine, 26*, 407–418.

Palácios, J. (2004). Mudança e Desenvolvimento Durante a Idade Adulta e a Velhice. In C. Coll, J. Palacios, & A. Marchesi (Eds.), *Desenvolvimento Psicológico e Educação Psicologia Evolutiva* (2nd ed., Vol. 1). Porto Alegre: Artmed.

Pate, R. R., Pratt, M., Blair, S. N., Haskell, W. L., Macera, C. A., Bouchard, C., et al. (1995). Physical activity and public health. A recommendation from the centers for disease control and prevention and the American College of Sports Medicine. *JAMA, 273*, 402–407.

Paterson, D., & Warburton, D. (2010). Physical activity and functional limitations in older adults: Asystematic review related to Canada's Physical Activity Guidelines. *International Journal of Behavioural Nutrition and Physical Activity, 7*, 1–22.

Paterson, D. H., Jones, G. R., & Rice, C. L. (2007). Le vieillissementetl' activité physique: Donnéessurlesquelles fonder des recommandations relatives à l'exercice à l'intention des adultes ages. *Applied Physiology, Nutrition and Metabolism, 32*, 75–121.

Reis J. (1995). O envelhecimento. In T. Geriátricos (Ed.), *1º tomo*. Lisboa: Prismédica.

Roach, S. S. (2003). *Introdução à Enfermagem Gerontológica*. Rio de Janeiro: Editora Guanabara Koogan.

Robertson, M., Campbell, A., Gardner, M., & Devlin, N. (2002). Preventing injuries in older people by preventing falls: meta-analysis of individual-level data. *Journal American Geriatrics, 50*, 905–911.

Robertson, M., Devlin, N., Gardner, M., & Campbell, A. (2001). Effectiveness and economic evaluation of a nurse delivered home exercise program to prevent falls. 1: Randomised controlled trial. *BMJ, 322*, 1–6.

Sandvik, L., Erikssen, J., Thaulow, E., Erikssen, G., Mundal, R., & Rodahl, K. (1993). Physical fitness as a predictor of mortality among healthy, middle-aged Norwegian men. *New England Journal of Medicine, 328*, 533–537.

Singh, M. A. (2002). Exercise comes of age: Rationale and recommendations for a geriatric exercise prescription. *Journal of Gerontology: Medical Sciences, 57*, M262–M282.

Slentz, C. A., Houmard, J. A., & Kraus, W. E. (2007). Modest exercise prevents the progressive disease associated with physical inactivity. *Exercise and Sports Science Reviews, 35*, 18–23.

Smilios, I., Pilianidis, T., Karamouzis, M., Parlavantzas, A., & Tokmakidis, S. P. (2007). Hormonal responses after a strength endurance resistance exercise protocol in young and elderly males. *International Journal of Sports Medicine, 28*, 401–406.

Sobolski, J., Kornitzer, M., De Backer, G., Dramaix, M., Abramowicz, M., Degre, S., et al. (1987). Protection against ischemic heart disease in the Belgian Physical Fitness Study: Physical fitness rather than physical activity? *American Journal of Epidemiology, 125*, 601–610.

Spirduso, W. W. (2005). *Physical Dimensions of Aging*. Champaign, IL: Human Kinetics.

Staab, A. S., & Hdges, L. C. (1997). *Enfermería gerontológica adaptación al proceso de envejecimiento*. México: McGraw—Hill Interamericana.

Steiner, R., Meyer, K., Lippuner, K., Schmid, J., Saner, H., & Hoppeler, H. (2004). Eccentric endurance training in subjects with coronary artery disease: A novel exercise paradigm in cardiac rehabilitation? *European Journal of Applied Physiology, 91*, 572–578.

Stella, F., Gobbi, S., Corazza, D. I., & Costa, J. L. R. (2002). Depressão no idoso: Diagnóstico, tratamento e benefícios da atividade física. *Motriz, 8*, 91–98.

Vicente, M. (2012). *Manual de enfermeira geriátrica*. Madrid: CTO Editorial.

Vieira, E. B., & Koenig, A. M. (2002). Avaliação Cognitiva. In E. V. Freitas, L. Py, A. L. Néri, F. A. Cançado, M. L. Gorzoni, & S. M. Rocha (Eds.), *Tratado de Geriatria e Gerontologia* (pp. 921–928). Rio de Janeiro: Guanabara Koogan.

Wieser, M., & Haber, P. (2007). The effects of systematic resistance training in the elderly. *International Journal of Sports Medicine, 28*, 59–65.

World Health Organization (1998). The role of physical activity in healthy ageing. Acedido em http://whqlibdoc.who.int/hq/1998/WHO_HPR_AHE_98.2.pdf.

Yassuda, M. S. (2002). Memória e envelhecimento saudável. In E. V. Freitas, L. Py, A. L. Neri, F. X. Cançado, M. Gorzoni, & S. Rocha (Eds.), *Tratado de geriatria e Gerontologia* (pp. 914–920). Rio de Janeiro: Guanabara Koogan.

Chapter 3
Activity Programmes for the Elderly

Gonçalo Nuno Figueiredo Dias, Micael Santos Couceiro
and Rui Mendes

Abstract A number of studies have described the benefits of regular physical exercise for the treatment and control of degenerative disorders. By recognizing the importance of physical activity for improving the quality of life for the elderly, the aim of this chapter is to provide a pedagogical and educational tool for the professionals who work with the senior population. Therefore, a physical activity programme was designed, which comprehends sticks, balls, hoops, resistance bands, chairs, among other methods. It is concluded that the physical activity programme for the elderly presupposes the inclusion of properly validated tests in order to evaluate this population. This way, it is important to carefully evaluate physical parameters such as cardio-respiratory capacity, muscular endurance, flexibility, agility and body composition. Finally, the prescription of physical activity for the elderly entails knowing/implies knowledge of age-related limitations, existing pathologies and individual changes arising from ageing.

Keywords Physical exercise · Elderly · Ageing · Health · Active life

3.1 Regular Physical Activity and Healthy Ageing

Several studies systematically link regular physical activity with healthy ageing, thus demonstrating that the quality of life of the elderly can be significantly improved by following physical activity programmes well suited to their psychomotor skills (e.g. Clark and Cotton 1998; Singh 2002; Jones and Rose 2005; Spirduso 2005; Irvine et al. 2013). Carrying out physical exercises is particularly valuable in the treatment and control of degenerative diseases, such as diabetes, osteoporosis and cardiovascular diseases (cf. Matsudo and Matsudo 1992; Smith 1993; Cassilhas et al. 2007; Smith et al. 2013).

Going down the rabbit hole, many researchers, namely from the field of gerontology, argue that ageing leads to biopsychosocial conditions (e.g. Clark and Cotton 1998; Meirelles 2000; Geis 2002; Dias et al. 2013, 2014, 2015):

(i) *Biological domain*:

1. Sensory impairments;
2. Loss of balance;
3. Postural changes;
4. Decreased pulmonary ventilation;
5. Decay in glucose and lipid metabolism;
6. Progressive strength loss, flexibility and muscle power;
7. Decreased motor coordination.

(ii) *Psychological domain*:

1. Decreased capacity to process information and to maintain the levels of concentration and attention;
2. Central nervous system deterioration in terms of perception, transmission and execution of nerve impulses.

(iii) *Social domain*:

1. Social depreciation;
2. Restricted social inclusion;
3. Social isolation;
4. Loss of social status.

In spite of this, the physical activity plays an important role in the active life of the elderly, especially for the recovery and treatment of chronic vascular diseases, hypertension and diabetes (see Hooke and Zoller 1992; Clark and Cotton 1998; Jones and Rose 2005; Dias et al. 2013). Clark and Cotton (1998), ACSM'S (2000), Geis (2002) and Irvine et al. (2013) have been demonstrating that the physical activity is crucial to overcome the previous conditions within all domains.

(i) *Biological domain*:

1. Increase the maximum volume of oxygen;
2. Increase the stroke volume which supports the cardiac output;
3. Decrease the resting heart rate, which increases the dilation of the heart and amplifies the stroke volume;
4. Maintenance of the motor coordination capacity;
5. Strengthening of the musculoskeletal system;
6. Increase the strength, endurance and muscle tone;
7. Prevention of osteoporosis;
8. Prevention of infections, as well as muscle and joint pains.

(ii) *Psychological domain*:

1. Optimization of cognitive functions;
2. Reduction of stress and anxiety;
3. Improvement of mood.

(iii) *Social domain*:
1. Establishing new relationships and friendships;
2. Social integration among peers;
3. Improvement of intergenerational activity.

Nevertheless, physical activity programmes designed for the elderly must be delivered by qualified professionals and should include the following features (see Geis 2002; Geis and Rubi 2003; Norman 2010):

1. Rewarding: it should be rewarding and appealing for the elderly;
2. Recreational: it should cover a playful and recreational component;
3. Motivational: it should motivate the elderly to achieve its objectives;
4. All-inclusive: it should integrate all the elderly population, while still considering their limitations;
5. Adaptive: it should be adapted to multiple possibilities and psychomotor needs of the elderly;
6. Socializing: it should have a social meaning and foster peer relationships.

For that reason, a physical activity programme aimed at the elderly needs to contemplate the changes resulting from the ageing process, focusing in the following objectives (Matsudo and Matsudo 1992; Singh 2002; Mcardle et al. 2003; Norman 2010):

1. Improve the overall strength and cardiovascular capacity;
2. Optimize the flexibility, balance and coordination;
3. Improve muscle strength and endurance;
4. Control the weight and nourishment;
5. Promote social contact and satisfaction;
6. Improve self-esteem and self-image;
7. Promote psychological well-being;
8. Encourage group dynamics.

While taking into account the following recommendations (see Matsudo and Matsudo 1992; Topp et al. 1993; Singh 2002):

1. Avoid fast and sudden movements;
2. The intensity of exercises should be low or moderate;
3. The exercises should not be applied whenever the elderly express pain;
4. The elderly can never be brought to exhaustion during exercise;
5. The application of anaerobic exercises is not recommended;
6. Isometric exercises are not advisable;
7. Exercises should preferably aim the large muscle groups, not just the located muscles;
8. Avoid exercises demanding sudden movements at the cervical level;
9. It is not recommended to exceed the maximum range of motion;
10. The use of drugs must be managed and planned;
11. Entertainment components, such as games, songs and traditional dances, can and should be integrated as a socialization and group dynamics enabler.

3.2 Physical Fitness Evaluation

It is important to evaluate the physical fitness of the elderly to measure his/her motor skills (Hooke and Zoller 1992; Geis 2002; Norman 2010). Hence, it is recommended that participants undergo a rigorous and thorough medical evaluation before starting any physical activity programme for this population (Geis 2002; Geis and Rubi 2003). It should also be noted that the term of responsibility, usually requested by certain sports facilities, such as gyms and swimming pools, does not prove the physical and mental strength of the elderly. Thus, this preliminary assessment is one of the many competences that qualified health professionals need to possess and should be carried out before prescribing any exercises.

Structuring a physical activity programme for the senior citizen must consider fully validated tests in order to thoughtfully evaluate this population (Baptista and Sardinha 2005). For Baptista and Sardinha (2005, p. 5), the Fullerton battery is an efficient tool to assess the functional physical fitness of people over 60 years old (Rikli and Jones 2001). This battery is useful to measure the musculoskeletal, cardio-respiratory and neurological abilities of the elderly, taking into account the evaluation of physical parameters, such as cardio-respiratory capacity, muscular strength, flexibility, agility and body composition. A brief description of the Fullerton battery is herein presented, highlighting some of the tests used to assess the functional physical fitness of the elderly (Baptista and Sardinha 2005; Dias et al. 2013):

1. *Stand up and sit down in the chair*: number of executions over 30 s without using the arms—allows to assess the strength and endurance of the lower limbs;
2. *Forearm Flexion*: number of executions over 30 s—allows to evaluate the strength and endurance of the upper limbs;
3. *Structure and weight*: allows to evaluate the body mass index;
4. *Sit and reach*: distance travelled by the hand towards the toes—allows to assess the flexibility of the hamstrings and lower back;
5. *Sit, walk and sit again*: time to rise from a chair, walk 2.44 m and return to the starting position—evaluates the speed, agility and balance;
6. *Reach behind the back*: minimum distance reached between the hands behind the back—allows to assess the flexibility of the shoulder;
7. *Walk*: travelled distance over six minutes—allows to evaluate the aerobic capacity;
8. *Step in place*: number of steps of the lower limbs without displacement over two minutes—allows to evaluate the aerobic capacity, as an alternative to the previous walk test;
9. *Stay with eyes closed and feet together*: standing position with a reduced base of support—allows to evaluate the usability of the proprioceptive information;

10. *Balance*: raise the dominant land leg and maintain balance for 20 s—allows to evaluate the unipedal balance;
11. *Postural reaction control*: the participant slowly supports his/her back in the evaluator's palm—allows the capability to effectively restore the balance after an unexpected disruption.

It is noteworthy that, given the characteristics of the aforementioned tests, the Fullerton battery shall be applied by qualified professionals.

3.3 Physical Activity Prescription

Prescribing physical activity for the elderly is something that entails knowing their limitations, any existing pathologies and the individual changes arising from ageing (Jacob 2007; Martins et al., 2012; Mendes et al., 2013, Silva 1993). Following the same insights proposed by several researchers, we herein present a set of recommendations to adequately plan and implement a physical activity programme targeted for this specific population (Clark and Cotton 1998; Geis 2002; Jones and Rose 2005; Norman 2010; ACSM'S 2000).

3.3.1 Structure

The structure of the physical activity session must contemplate three distinct phases: (cf., Geis 2002; Singh 2002; Mcardle et al. 2003; Norman 2010):

1. Initial phase: joint and organic mobilization;
2. Central phase: learning and/or performance of physical exercise;
3. Final Part: return to a normal resting state.

3.3.2 Frequency

The frequency of the physical activity sessions must be, in average, two or three times a week, with a resting interval of 24–48 h between sessions so as to ensure the physical recovery of the elderly (Geis 2002; Singh 2002; Mcardle et al. 2003; Norman 2010).

3.3.3 Duration

The duration of each physical activity session must vary between 40 and 60 min of physical activity (Singh 2002; Mcardle et al. 2003; Norman 2010).

3.3.4 Intensity

Given that the physical activity aimed at the elderly is predominantly aerobic with a low to moderate intensity, the proposed exercises must lead to 60–70% of the maximum heart rate (Singh 2002; Mcardle et al. 2003; Norman 2010).

3.3.5 Repetitions per Exercise

Each exercise should not exceed 10–15 repetitions, respecting a recovery time of 2–3 min between the same motion series (Singh 2002; Mcardle et al. 2003; Norman 2010).

3.3.6 Technical Indications

At the beginning of the session, the elderly shall perform flexibility exercises to prevent injury in the joints, tendons, ligaments and muscles (Norman 2010). At the end of the session, the elderly shall perform stretches which may contemplate multiple 15–30 s exercises (Norman 2010).

3.4 General Exercises

This section depicts several exercises, adequately represented with real pictures, which allow to illustrate, in detail, the methodologies presented in the contemporary literature under the topic of active ageing and physical activity for the elderly. It is noteworthy that all exercises should be carried out slowly, without losing balance, and within the range of motion of the elderly. Additionally, for the sake of simplicity, common terms were used to describe the exercises, including "arms" (upper limbs), "legs" (lower limbs) and so on (Dias et al. 2013, 2014, 2015).

3.4 General Exercises

3.4.1 Stick

Exercise #1

| Hold the stick at shoulder height with the palms facing down | Raise the arms over the head |

Objective: Promote the movement of the shoulder girdle
Organization: Individual
Material: Stick

Exercise #2

(continued)

(continued)

Hold the stick at shoulder height with the palms facing down	Driving the stick and the trunk from one side to the other

Objective: Promote the movement of the shoulder girdle and strengthen the trunk muscles

Organization: Individual

Material: Stick

Exercise #3

Raise the arms over the head with the palms facing up	Bending the torso from one side to the other

Objective: Promote the movement of the shoulder girdle and strengthen the trunk muscles

Organization: Individual

Material: Stick

3.4 General Exercises

Exercise #4

	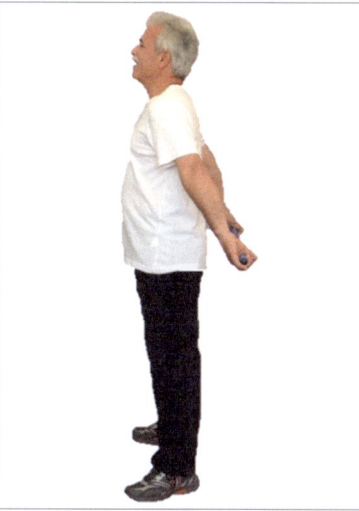
Hold the stick behind the pelvis with the palms facing up	Drive the stick away and near the pelvic girdle

Objective: Promote the movement of the shoulder girdle and strengthen the triceps
Organization: Individual
Material: Stick

Exercise #5

Raise the arms over the head with the palms facing up	Bend the torso forward and back to the starting position

(continued)

(continued)

Objective: Promote movement of the trunk muscles
Organization: Individual
Material: Stick

Exercise #6

| Hold the stick at shoulder height with the palms facing down | Bend and stretch the legs |

Objective: Promote the movement of the joints and muscles of the legs
Organization: Individual
Material: Stick

Exercise #7

(continued)

3.4 General Exercises

(continued)

Hold the stick at shoulder height with the palms facing down	Raise and lower the wrists
Objective: Promote movement of the wrists' joints	
Organization: Individual	
Material: Stick	

3.4.2 Ball

Exercise #1	

Hold the ball at chest height	Throw the ball upwards and grab it with both hands
Objective: Promote the visuomotor skills and balance	
Organization: Individual	
Material: Ball	

Exercise #2

Hold the ball at chest height	Throw the ball upwards and clap your hands once, grabbing it with both hands afterwards

Objective: Promote visuomotor skills, reaction, rhythm and balance

Organization: Individual

Material: Ball

Exercise #3

(continued)

3.4 General Exercises

(continued)

Hold the ball at pelvis height	Pass the ball behind his back, from one hand to the other, around the pelvic girdle

Objective: Promote visuomotor skills, rhythm and balance

Organization: Individual

Material: Ball

Exercise #4

Hold the ball at shoulder height	Raise the arms above the head and squeeze the ball with both hands

Objective: Promote the movement of the shoulder girdle and joint hands

Organization: Individual

Material: Ball

Exercise #5

Hold the ball in the knees	Bend and stretch the legs without dropping the ball, while keeping the hands resting on the pelvic girdle

Objective: Promote the movement of the legs

Organization: Individual

Material: Ball

Exercise #6

(continued)

3.4 General Exercises

(continued)

Hold the ball at shoulder height	Open and close both hands without dropping the ball
Objective: Promote the movement of the hand muscles and joints	
Organization: Individual	
Material: Ball	

Exercise #7

Hold your arms laterally stretched at the shoulder height	Pass the ball over the head from one hand to the other, without throwing it, keeping the arms stretched
Objective: Promote the movement of the arms and shoulder girdle	
Organization: Individual	
Material: Ball	

3.4.3 Hoop

Exercise #1

Hold the hoop and keep the upper limbs at the shoulder width	Throw the hoop upwards and grab it with the same hand

Objective: Promote visuomotor skills, rhythm and balance

Organization: Individual

Material: Hoop

3.4 General Exercises

Exercise #2

Hold the hoop down and keep the upper limbs at the shoulder width	Throw the hoop upwards and grab it with the other hand

Objective: Promote visuomotor skills, rhythm and balance

Organization: Individual

Material: Hoop

Exercise #3

 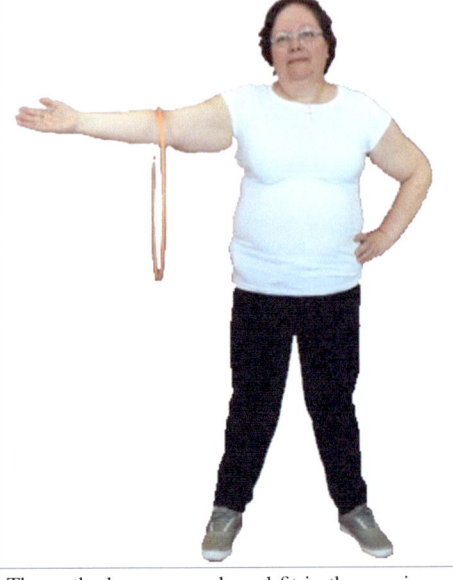

Hold the hoop down and keep the upper limbs at the shoulder width	Throw the hoop upwards and fit in the arm in mid-air

Objective: Promote visuomotor skills, rhythm and balance

Organization: Individual

Material: Hoop

Exercise #4

(continued)

3.4 General Exercises

(continued)

Hold the hoop at shoulder height	Simulate the driving car motion with both hands, crossing the arms while stretched

Objective: Promote the movement of the shoulder girdle and arm muscles
Organization: Individual
Material: Hoop

Exercise #5

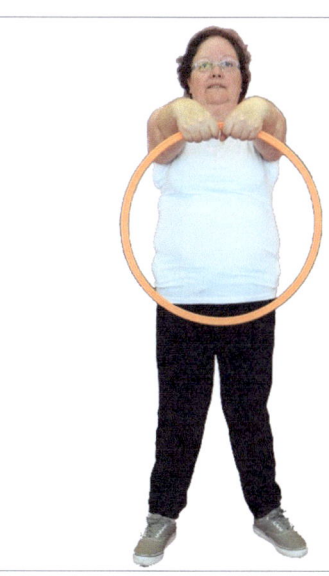

Hold the hoop at shoulder height	Grab the hoop with both hands and rotate it up and down, using only the pulses

Objective: Promote movement of the wrists
Organization: Individual
Material: Hoop

Exercise #6

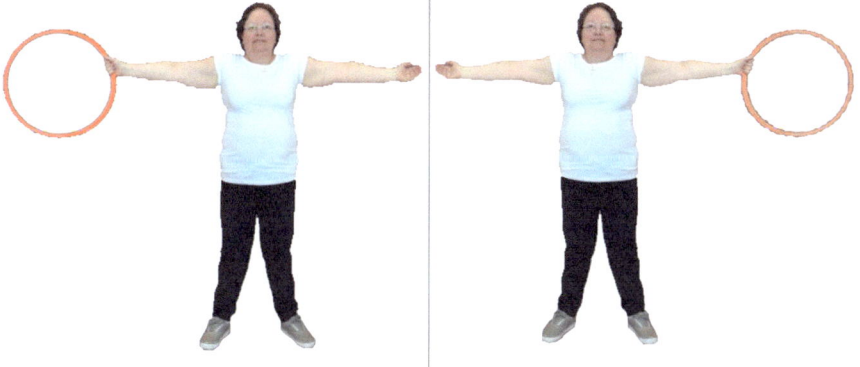

(continued)

(continued)

Hold your arms laterally stretched at the shoulder height	Pass the hoop over the head from one hand to the other, without throwing it, keeping the arms stretched

Objective: Promote the movement of the shoulder girdle and arm muscles

Organization: Individual

Material: Hoop

Exercise #7

Hold the hoop down and keep the upper limps at the shoulder width	Pass the hoop around the pelvic girdle

Objective: Promote the movement of the shoulder girdle and arm muscles

Organization: Individual

Material: Hoop

3.4 General Exercises

Exercise #8

| Hold the hoop at shoulder height | Flex and stretch the arms while holding the hoop |

Objective: Promote the movement of the shoulder girdle, elbow and arm muscles
Organization: Individual
Material: Hoop

Exercise #9

| Hold the hoop down and keep the upper limbs at the shoulder width | Raise the hoop up to shoulder height, keeping the arm laterally stretched |

Objective: Promote the movement of the shoulder girdle and arm muscles
Organization: Individual
Material: Hoop

3.4.4 Resistance Band (Part 1)

Exercise #1

| Hold the band at shoulder height | Hold the band with both hands and stretch it sideways |

Objective: Promote the movement of the shoulder girdle and arm muscles

Organization: Individual

Material: Resistance band

Exercise #2

(continued)

3.4 General Exercises

(continued)

Hold the band at chest height	Hold the band with one hand facing down, while the other pulls the elastic up (flip over the band after 8 reps)

Objective: Promote the movement of the shoulder girdle and arm muscles

Organization: Individual

Material: Resistance band

Exercise #3

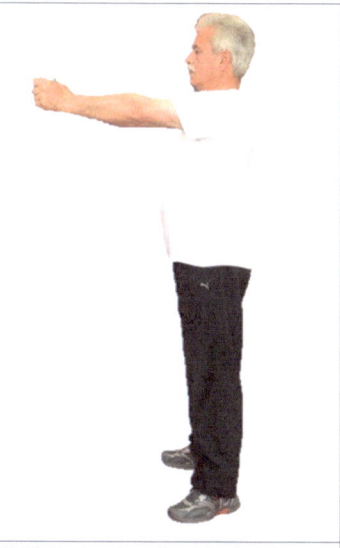 

Hold the band at shoulder height	Hold the band with both hands and turn the handles

Objective: Promote the movement of the wrists and the hands muscles

Organization: Individual

Material: Resistance band

Exercise #4

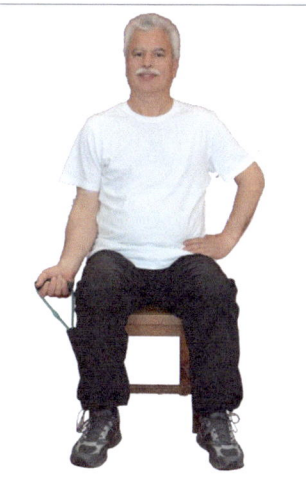 

(continued)

(continued)

Sit on a chair and with the chair leg holding the band	Bend and stretch the arm onwards with the palm facing up (flip over the rubber band to the other arm after 8 reps)
Objective: Promote the biceps toning	
Organization: Individual	
Material: Resistance band and chair	

Exercise #5

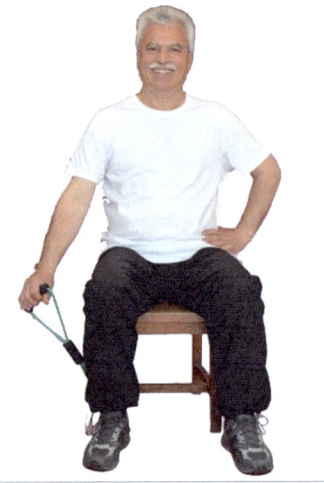 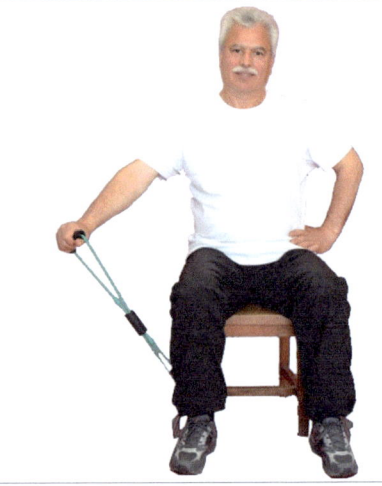

Sit on a chair and with the chair leg holding the band	Bend and stretch the arm sideways with the palm facing down (flip over the rubber band to the other arm after 8 reps)
Objective: Promote the movement of the shoulder girdle and arm muscles	
Organization: Individual	
Material: Resistance band and chair	

3.4 General Exercises

Exercise #6

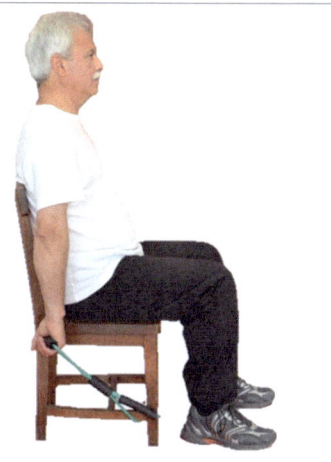	
Sit on a chair and with the chair leg holding the band	Bend and stretch the arm backwards (flip over the rubber band to the other arm after 8 reps)

Objective: Promote the triceps toning

Organization: Individual

Material: Resistance band and chair

Exercise #7

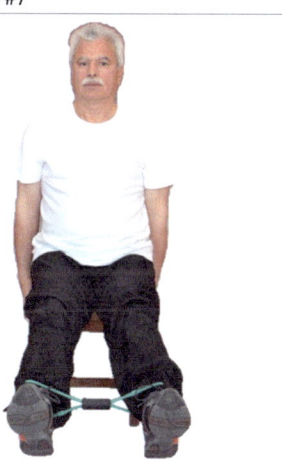	
Sit on a chair, with the hands holding it and with the band in the ankles	Abduct and adduct the legs while keeping them stretched

Objective: Promote the movement of the legs joints and muscles

Organization: Individual

Material: Resistance band and chair

Exercise #8

| Sit on a chair, with the hands holding it and with the band in the ankles | Alternatively raise and lower the legs while keeping them stretched |

Objective: Promote the movement of the legs joints and muscles

Organization: Individual

Material: Resistance band and chair

Exercise #9

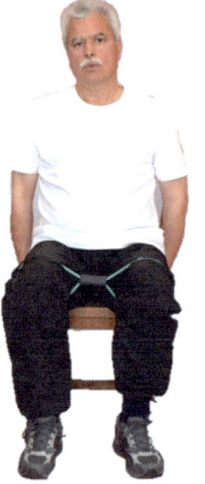 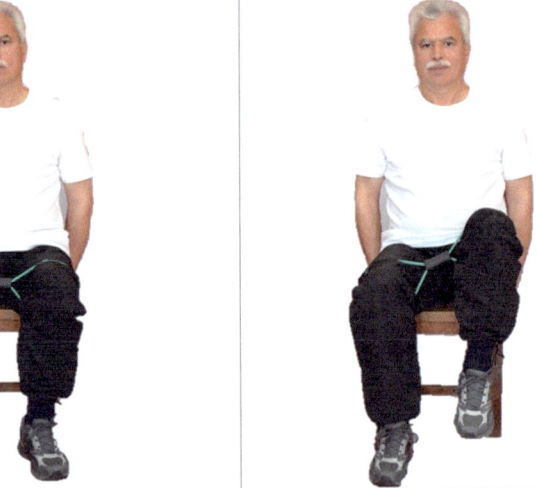

| Sit on a chair, with the hands holding it and with the band in the thighs | Alternatively raise and lower the legs while keeping them bended |

Objective: Promote the movement of the legs joints and muscles

Organization: Individual

Material: Resistance band and chair

3.4.5 Resistance Band (Part 2)

Exercise #1

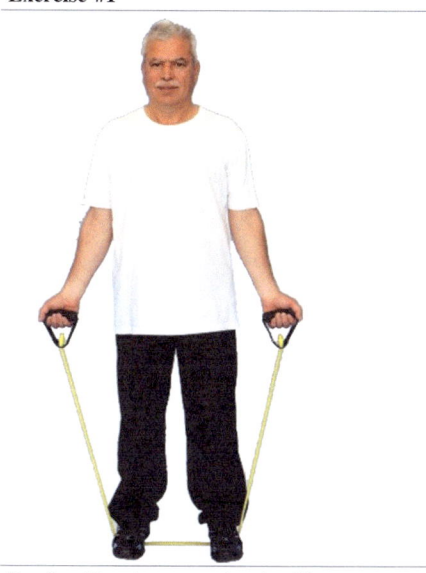
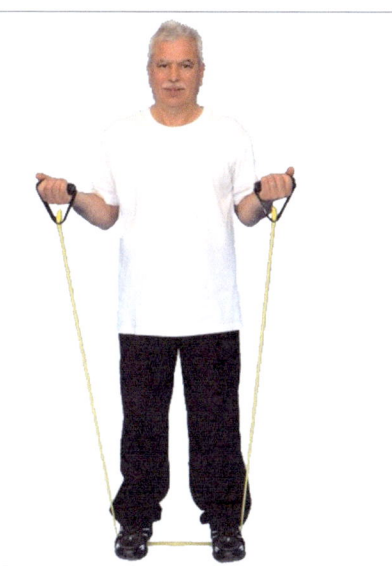

Trunk still and arms at the shoulder width, with the feet holding the band against the floor	Bend and stretch the arms

Objective: Promote the biceps toning

Organization: Individual

Material: Resistance band

Exercise #2

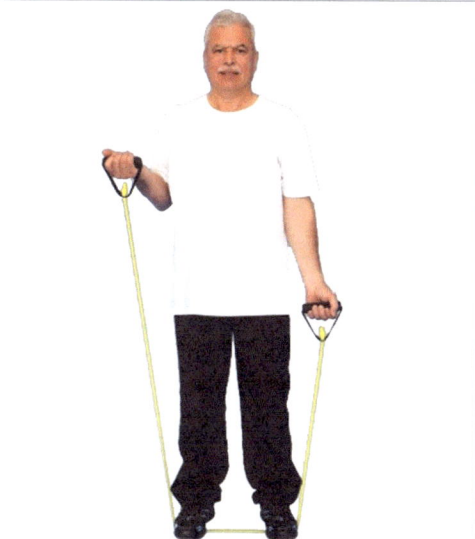
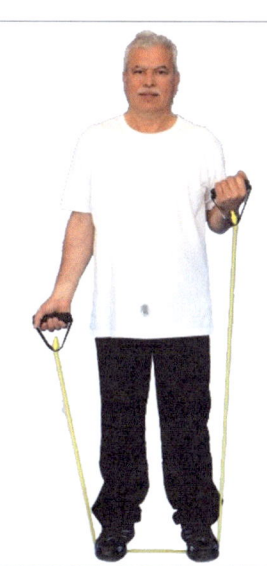

(continued)

(continued)

Trunk still and arms at the shoulder width, with the feet holding the band against the floor	Alternatively bend and stretch the arms

Objective: Promote the biceps toning

Organization: Individual

Material: Resistance band

Exercise #3

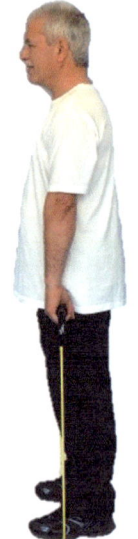

Trunk still and arms at the shoulder width, with the feet holding the band against the floor	Stretch the arms backwards

Objective: Promote the triceps toning

Organization: Individual

Material: Resistance band

3.4 General Exercises

Exercise #4

| Trunk still and arms at the shoulder width, with the feet holding the band against the floor | Alternatively stretch the arms backwards |

Objective: Promote the triceps toning

Organization: Individual

Material: Resistance band

Exercise #5

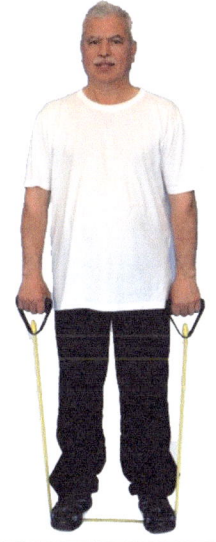
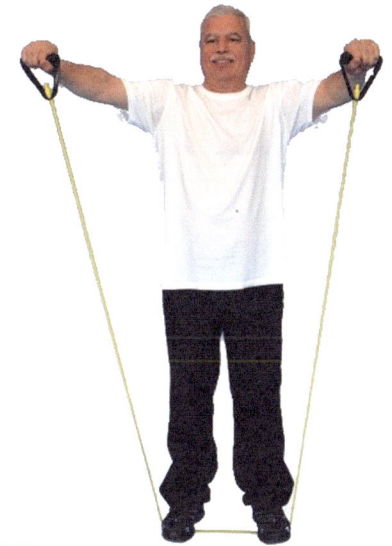

(continued)

(continued)

Trunk still and arms at the shoulder width, with the feet holding the band against the floor	Stretch the arms to shoulder height
Objective: Promote the movement of the shoulder girdle and arm muscles	
Organization: Individual	
Material: Resistance band	

Exercise #6

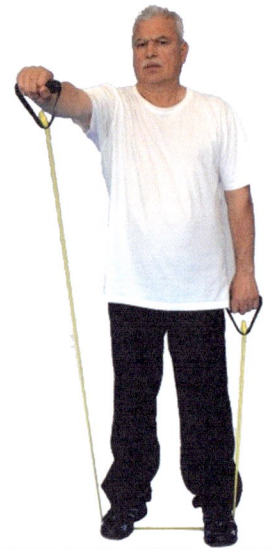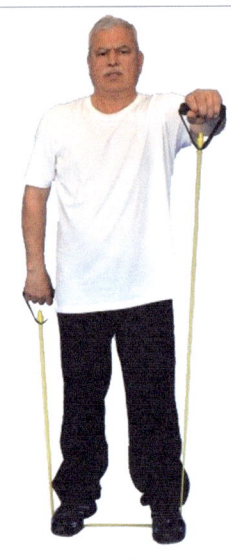

Trunk still and arms at the shoulder width, with the feet holding the band against the floor	Alternatively stretch each arm to shoulder height
Objective: Promote the movement of the shoulder girdle and arm muscles	
Organization: Individual	
Material: Resistance band	

3.4 General Exercises

Exercise #7

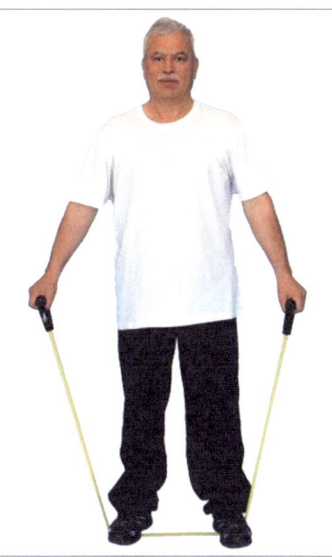

Trunk still and arms at the shoulder width, with the feet holding the band against the floor	Laterally stretch the arms to shoulder height

Objective: Promote the movement of the shoulder girdle and arm muscles

Organization: Individual

Material: Resistance band

Exercise #8

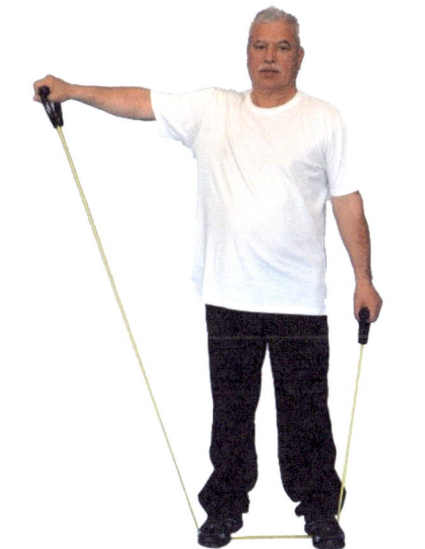
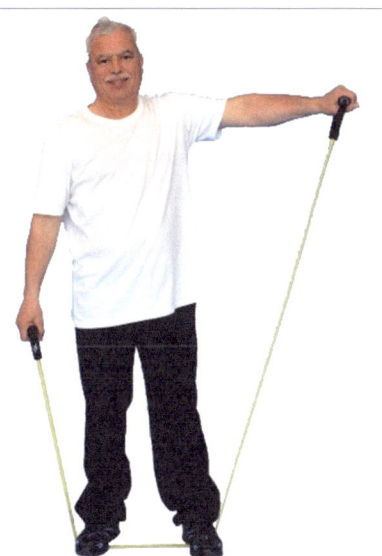

(continued)

Trunk still and arms at the shoulder width, with the feet holding the band against the floor	Alternatively and laterally stretch the arms to shoulder height |//
Objective: Promote the movement of the shoulder girdle and arm muscles	
Organization: Individual	
Material: Resistance band	

Exercise #9

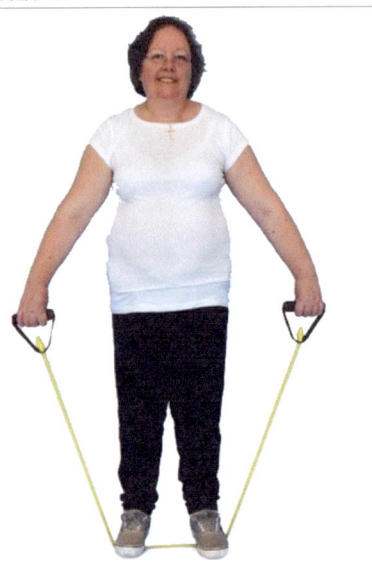 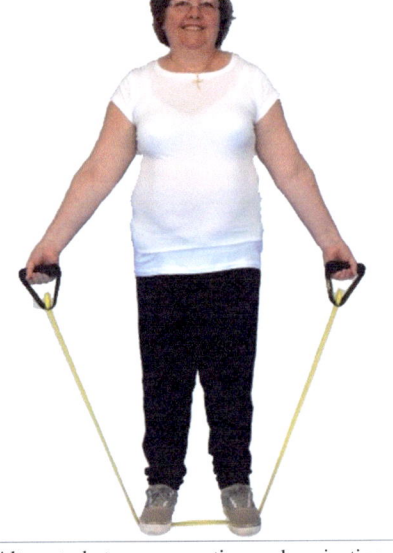

Trunk still and arms at the shoulder width, with the feet holding the band against the floor	Alternate between pronation and supination while keeping the arms laterally stretched
Objective: Promote the movement of the forearms, wrists and shoulder girdle	
Organization: Individual	
Material: Resistance band	

3.4 General Exercises

Exercise #10

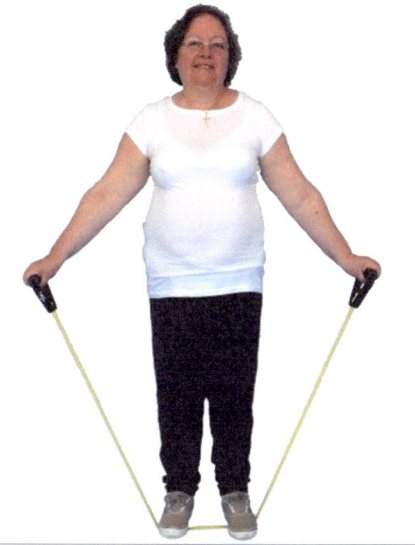 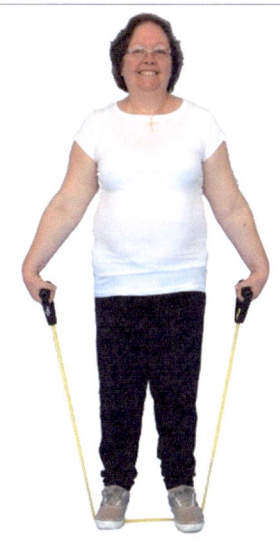

| Trunk still and arms at the shoulder width, with the feet holding the band against the floor | Rotate the band out and in, away and towards the trunk, while keeping the arms laterally stretched |

Objective: Promote the movement of the forearms and wrists

Organization: Individual

Material: Resistance band

3.4.5.1 Chair

Exercise #1

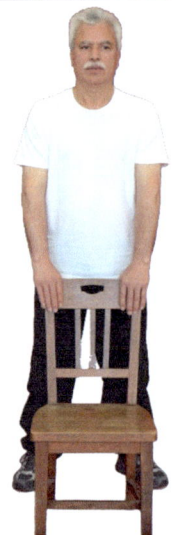 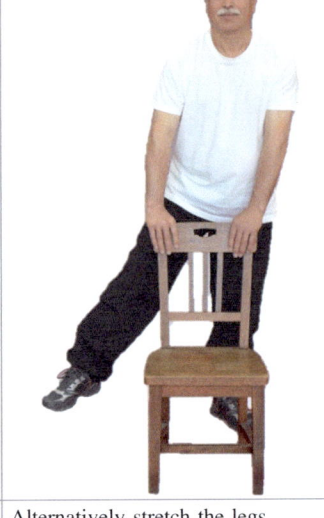

Stand behind the chair, with hands shoulder width apart and resting on the back of the chair	Alternatively stretch the legs sideways without losing balance

Objective: Promote the movement of the legs joints and muscles

Organization: Individual

Material: Chair

Exercise #2

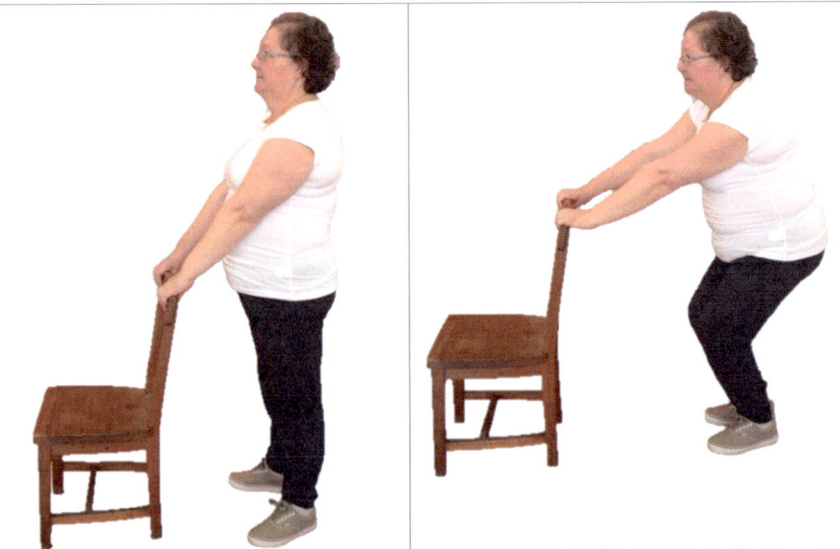

(continued)

3.4 General Exercises

(continued)

Stand behind the chair, with hands shoulder width apart and resting on the back of the chair	Bend and stretch the legs

Objective: Promote the movement of the legs joints and muscles

Organization: Individual

Material: Chair

Exercise #3

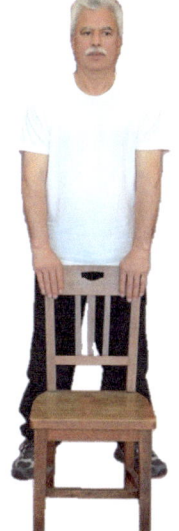

Stand behind the chair, with hands shoulder width apart and resting on the back of the chair	Alternatively raise and bend the legs to pelvic height

Objective: Promote the movement of the legs joints and muscles

Organization: Individual

Material: Chair

Exercise #4

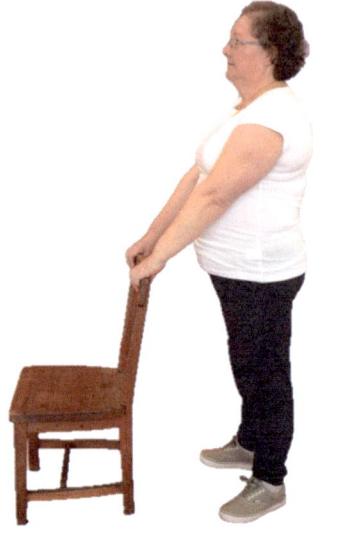 

Stand behind the chair, with hands shoulder width apart and resting on the back of the chair	Alternatively raise the heels at the rear

Objective: Promote the movement of the legs joints and muscles

Organization: Individual

Material: Chair

Exercise #5

(continued)

3.4 General Exercises

(continued)

Stand behind the chair, with hands shoulder width apart and resting on the back of the chair	Swing one leg forward and backward while supporting the body weight on the other
Objective: Promote the movement of the legs joints and muscles	
Organization: Individual	
Material: Chair	

Exercise #6

 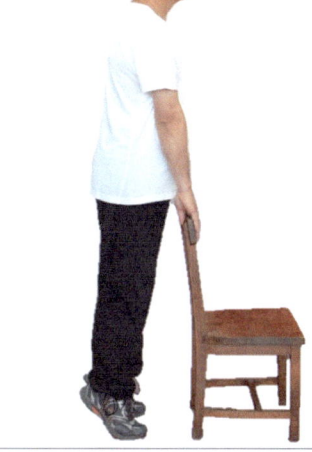

Stand behind the chair, with hands shoulder width apart and resting on the back of the chair	Raise and lower both heels simultaneously
Objective: Promote the movement of the tarsal joint and legs muscles	
Organization: Individual	
Material: Chair	

Exercise #7

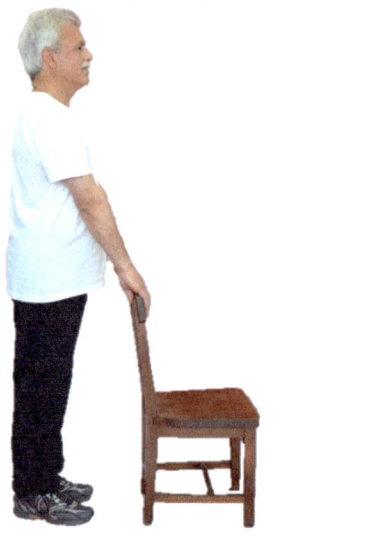

Stand behind the chair, with hands shoulder width apart and resting on the back of the chair	Alternatively raise and lower the heels

Objective: Promote the movement of the tarsal joint and legs muscles

Organization: Individual

Material: Chair

Exercise #8

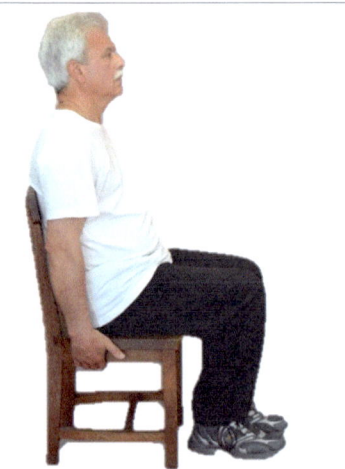

Sit on a chair, with the hands holding it	Simultaneously raise and lower both legs horizontally

Objective: Promote the movement of the legs joints and muscles

Organization: Individual

Material: Chair

3.4 General Exercises

Exercise #9

Sit on a chair, with the hands holding it	Alternatively raise and lower both legs horizontally

Objective: Promote the movement of the legs joints and muscles
Organization: Individual
Material: Chair

Exercise #10

Sit on a chair, with the hands holding it	Raise and maintain the legs stretched while simultaneously bending and stretching the feet

Objective: Promote the movement of the tarsal joint
Organization: Individual
Material: Chair

3.5 Strength Exercises

Strength training needs an additional care and attention so as to ensure the physical integrity of the elderly (Jones and Rose 2005; Norman 2010). Accordingly, the following procedures are of the essence to delineate an individual plan of strength exercises:

- Assess the body constitution and get the lean body mass, fat mass and body mass index;
- Analyse possible postural deviations that may constrain the use of free weight (dumb-bell);
- Assess the flexibility;
- Assess the body strength in various sectors, in order to design the optimal workload;
- Get detailed clinical history, by carrying out a diagnosis of potential health problems or limitations.

One must define the loads and define the amount of sets and repetitions for each individual plan and select the most appropriate exercises. The literature recommends carrying 2 or 3 sets of 8–10 repetitions each (Jones and Rose 2005; Norman 2010). The exercises should be carried out slowly, at low intensity, gradually increasing the load depending on the physical capabilities of the elderly. Generally, the most common load is set around 80% of one-repetition-maximum, which is equivalent to 8 repetitions (Clark and Cotton 1998; Jones and Rose 2005; Norman 2010). A 2–3-min rest between series is also recommended. In addition, exercises that might endanger the physical integrity of the elderly should be avoided, namely bench press, deadlift, squat and other free-weight exercises.

It is noteworthy that, considering the limitations inherent to the elderly population, all exercises require the supervision of a technical expert that may provide technical information, postural corrections and any other relevant assistance to efficiently carry out the strength exercises (Clark and Cotton 1998; Jones and Rose 2005; Norman 2010). The following checklist should be taken into account during the strength exercises:

- Body posture and breath control are of the utmost importance;
- Preferably focus in the major muscle groups over the local muscles;
- Isometric exercises are not advisable;
- It is not recommended to exceed the maximum range of motion;
- It is important to make a good joint and organic mobilization at the beginning of each session, as well as stretching at the end of the activity (Dias and Mendes 2013, 2014).

3.5 Strength Exercises

3.5.1 *Dumb-bells*

Exercise #1

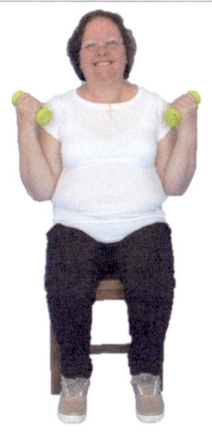 

Sit on a chair with a dumb-bell on each hand	Stretch and bend the arms

Objective: Promote the biceps toning
Organization: Individual
Material: Dumb-bells (1 kg each) and chair

Exercise #2

Sit on a chair with a dumb-bell on each hand	Alternatively stretch and bend the arms

Objective: Promote the biceps toning
Organization: Individual
Material: Dumb-bells (1 kg each) and chair

Exercise #3

Sit on a chair with a dumb-bell on each hand	Alternatively stretch the arms backwards

Objective: Promote the biceps toning
Organization: Individual
Material: Dumb-bells (1 kg each) and chair

Exercise #4

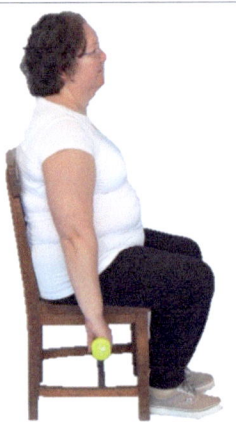 

Sit on a chair with a dumb-bell on each hand	Stretch the arms backwards

Objective: Promote the biceps toning
Organization: Individual
Material: Dumb-bells (1 kg each) and chair

3.5 Strength Exercises

Exercise #5

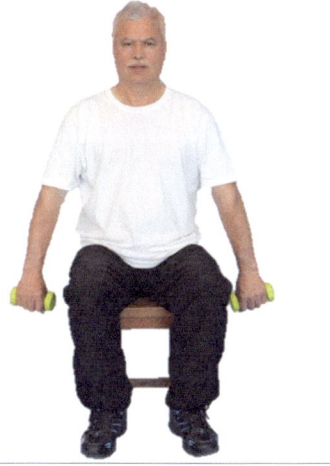 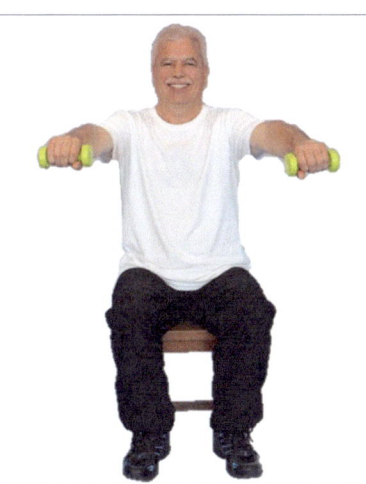

Sit on a chair with a dumb-bell on each hand	Stretch the arms at shoulder girdle height

Objective: Promote the movement of the shoulder girdle

Organization: Individual

Material: Dumb-bells (1 kg each) and chair

Exercise #6

 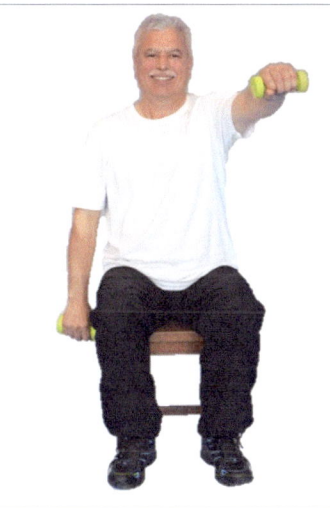

Sit on a chair with a dumb-bell on each hand	Alternatively stretch the arms at shoulder girdle height

Objective: Promote the movement of the shoulder girdle

Organization: Individual

Material: Dumb-bells (1 kg each) and chair

78 3 Activity Programmes for the Elderly

Exercise #7

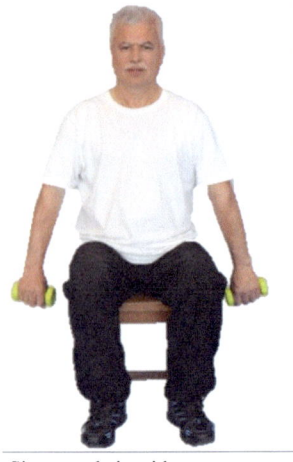 

Sit on a chair with a dumb-bell on each hand	Laterally stretch the arms at shoulder girdle height

Objective: Promote the movement of the shoulder girdle

Organization: Individual

Material: Dumb-bells (1 kg each) and chair

Exercise #8

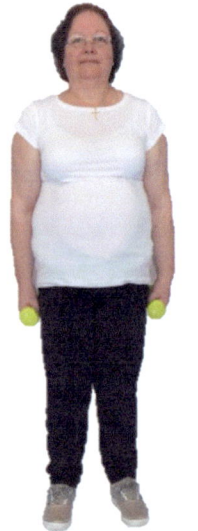 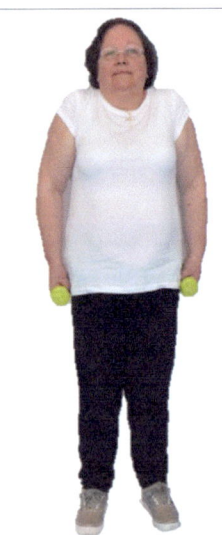

Hold a dumb-bell on each hand	Pull the shoulders up and down

Objective: Promote the movement of the shoulder girdle

Organization: Individual

Material: Dumb-bells (1 kg each)

3.5 Strength Exercises

Exercise #9

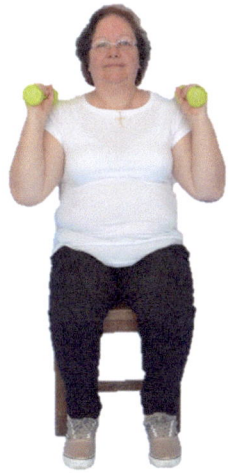 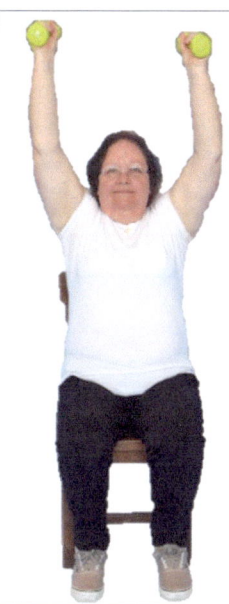

| Sit on a chair with a dumb-bell on each hand and at shoulder girdle height | Stretch and bend the arms |

Objective: Promote the movement of the shoulder girdle

Organization: Individual

Material: Dumb-bells (1 kg each) and chair

Exercise #7

(continued)

(continued)

Sit on a chair with a dumb-bell on each hand and supporting the wrists on the knees with the palms up (in supination)	Raise and lower the weights using only the wrists
Objective: Promote the movement of the forearm	
Organization: Individual	
Material: Dumb-bells (1 kg each) and chair	

Exercise #7

Hold a dumb-bell on each hand with the arms laterally extended	Rotate the pulses inward and outward relative to the body
Objective: Promote the movement of the forearms and wrists	
Organization: Individual	
Material: Dumb-bells (1 kg each)	

3.5 Strength Exercises

3.5.2 Neoprene Ankle Weights

Exercise #1	
	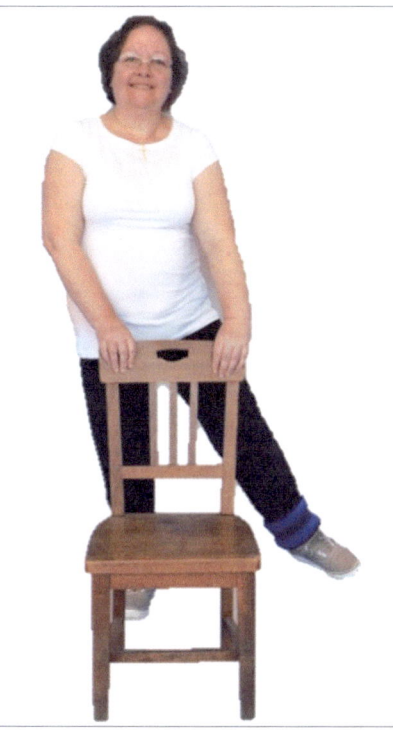
Stand behind the chair, with hands shoulder width apart and resting on the back of the chair	With neoprene ankle weights, alternatively stretch the legs sideways without losing balance
Objective: Promote the movement of the legs joints and muscles	
Organization: Individual	
Material: Neoprene ankle weights (1 kg each) and chair	

Exercise #2

Stand behind the chair, with hands shoulder width apart and resting on the back of the chair	With neoprene ankle weights, alternatively raise and bend the legs to pelvic height

Objective: Promote the movement of the legs joints and muscles

Organization: Individual

Material: Neoprene ankle weights (1 kg each) and chair

Exercise #3

(continued)

3.5 Strength Exercises

(continued)

Stand behind the chair, with hands shoulder width apart and resting on the back of the chair	With neoprene ankle weights, alternatively raise the heels at the rear
Objective: Promote the movement of the legs joints and muscles	
Organization: Individual	
Material: Neoprene ankle weights (1 kg each) and chair	

Exercise #4

Stand behind the chair, with hands shoulder width apart and resting on the back of the chair	With neoprene ankle weights, raise and lower both heels simultaneously
Objective: Promote the movement of the tarsal joint and legs muscles	
Organization: Individual	
Material: Neoprene ankle weights (1 kg each) and chair	

Exercise #5

Stand behind the chair, with hands shoulder width apart and resting on the back of the chair	With neoprene ankle weights, alternatively raise and lower the heels

Objective: Promote the movement of the tarsal joint and legs muscles

Organization: Individual

Material: Neoprene ankle weights (1 kg each) and chair

3.5 Strength Exercises

Exercise #6

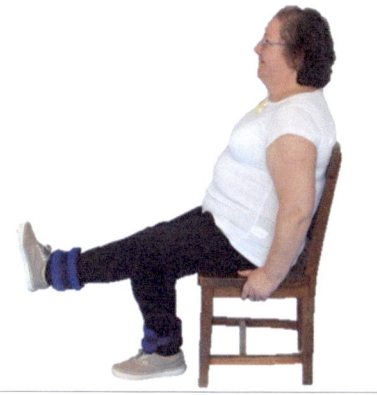 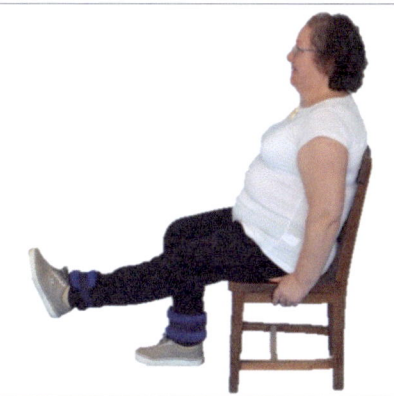

Sit on a chair, with the hands holding it	With neoprene ankle weights, alternatively raise and lower both legs horizontally

Objective: Promote the movement of the legs joints and muscles

Organization: Individual

Material: Neoprene ankle weights (1 kg each) and chair

Exercise #7

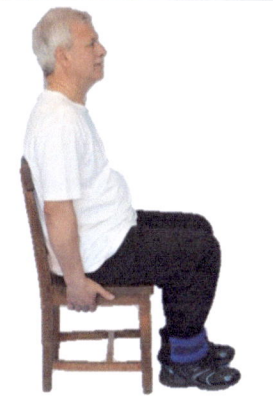 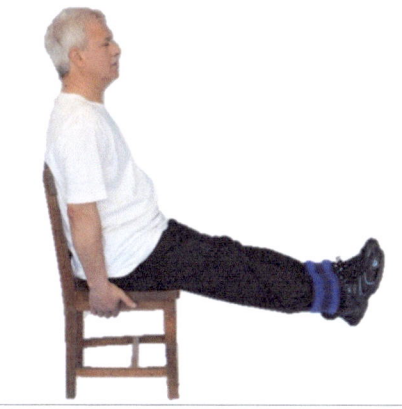

Sit on a chair, with the hands holding it	With neoprene ankle weights, simultaneously raise and lower both legs horizontally

Objective: Promote the movement of the legs joints and muscles

Organization: Individual

Material: Neoprene ankle weights (1 kg each) and chair

3.6 Partner Exercises

3.6.1 Body Strengthening

Exercise #1

While standing up, support one of the hands on the shoulders of each other and stretch the opposite leg sideways

Objective: Stimulate group dynamics, motor coordination and peer relations

Organization: In pairs

Material: None

3.6 Partner Exercises

Exercise #2

While standing up, support one of the hands on the shoulders of each other and raise and bend the opposite leg to pelvic height

Objective: Stimulate group dynamics, motor coordination and peer relations

Organization: In pairs

Material: None

Exercise #3

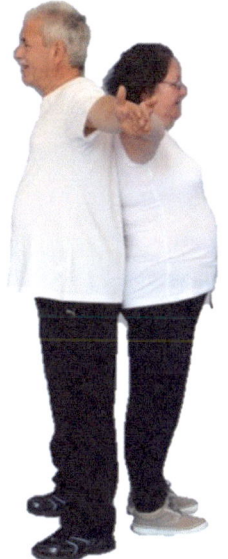

(continued)

(continued)

While standing up, with the back to each other and the arms laterally stretched, raise and lower the arms simultaneously
Objective: Stimulate group dynamics, motor coordination and peer relations
Organization: In pairs
Material: None

Exercise #4

While standing up, with the back to each other and the arms straight up, pass on the ball
Objective: Stimulate group dynamics, motor coordination and peer relations
Organization: In pairs
Material: Sponge ball

3.6 Partner Exercises

Exercise #5

While standing up, with the back to each other, pass on the ball laterally at the pelvic girdle height

Objective: Stimulate group dynamics, motor coordination and peer relations

Organization: In pairs

Material: Sponge ball

Exercise #6

(continued)

(continued)

While standing up, facing one another, raise and lower the arms simultaneously while holding two sticks

Objective: Stimulate group dynamics, motor coordination and peer relations

Organization: In pairs

Material: Two sticks

Exercise #7

(continued)

3.6 Partner Exercises

(continued)

While standing up, facing one another, raise and lower the arms alternatively while holding two sticks

Objective: Stimulate group dynamics, motor coordination and peer relations

Organization: In pairs

Material: Two sticks

Exercise #8

(continued)

(continued)

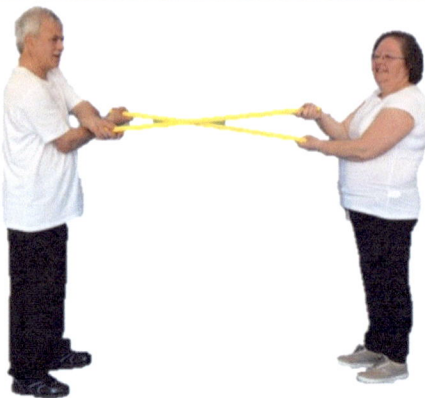

While standing up, facing one another, cross the arms while holding two sticks
Objective: Stimulate group dynamics, motor coordination and peer relations
Organization: In pairs
Material: Two sticks

Exercise #9

(continued)

(continued)

While standing up, facing one another and the arms straight up, stretch and bend the legs while holding two sticks
Objective: Stimulate group dynamics, motor coordination and peer relations
Organization: In pairs
Material: Two sticks

3.6.2 Body Language

Exercise #1

(continued)

(continued)

While standing up, hold hands and follow the movements proposed by the partner (switch after 4 exercises)
Examples of exercises: hold circles, stretch and bend, raise and lower, among others
Objective: Stimulate group dynamics, motor coordination and peer relations
Organization: In pairs
Material: None

Exercise #2

While standing up, imitate the exercises proposed by the partner (switch after 4 exercises)
Examples of exercises: Simulate the way birds flap their wings
Objective: Stimulate group dynamics, motor coordination and peer relations
Organization: In pairs
Material: None

3.6 Partner Exercises

Exercise #3

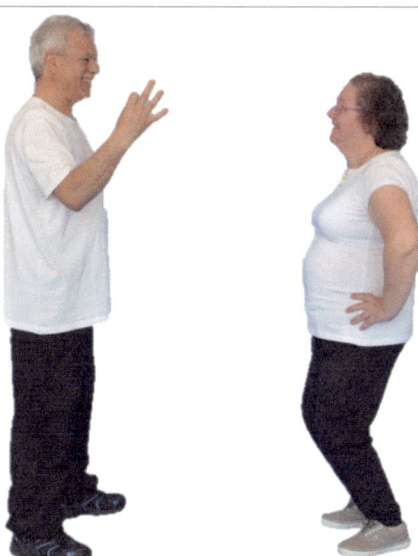

While standing up, follow the instructions given by the partner in the form of a number or number of clapping hands (switch after 4 exercises)

Examples of exercises: number (1) walk on space; number (2) open and close the hands; number (3) stretch and bend the legs five times; number (4) clap the hands; number (5) simulate playing the piano, among other moves

Objective: Stimulate group dynamics, motor coordination and peer relations

Organization: In pairs

Material: None

Exercise #4

(continued)

(continued)

While standing up, hold an A4 sheet of paper in each hand and simultaneously move them at the sound of the music played by the trainer (e.g. classical music, folklore, jazz, etc.)

Objective: Stimulate group dynamics, motor coordination and peer relations

Organization: In pairs

Material: Two A4 sheets of paper and stereo

Exercise #5

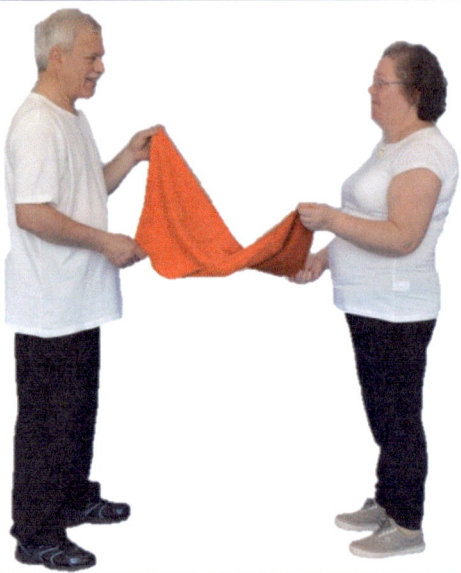

While standing up, hold the ends of a sheet to simulate the motion of waves with the arms (the legs can also follow the motion of waves)

Examples of exercises: if the sea is rough, the sheet should be shaken with irregular and fast movements representing tidal waves, while if the sea is calm, the sheet should be shaken with slow and gentle movements

Objective: Stimulate group dynamics, motor coordination and peer relations

Organization: In pairs

Material: Sheet

3.6 Partner Exercises

Exercise #6

While standing up, hold the extremities of two sticks with the forefingers or palms and imitate the movements of your colleague (switch after 4 exercises)
Examples of exercises: zoom in and out the stick from your body

Objective: Stimulate group dynamics, motor coordination and peer relations

Organization: In pairs

Material: Two sticks

Exercise #7

(continued)

(continued)

While standing up, 3 metres apart from each other, throw the ball 6 consecutive times each time the trainer tells you to do so
Objective: Stimulate group dynamics, motor coordination and peer relations
Organization: In pairs
Material: Sponge ball

3.7 Return to Resting State

To return to resting state, seniors who present ambulatory abilities can perform the exercises in a standing position, so the chair is not mandatory. Moreover, it is recommended that each stretching exercise lasts between 10 and 15 s each, with a maximum duration of 30 s.

Exercise #1

Interlace the fingers, point the palms facing forward and stretch the arms
Objective: Reduce muscle and joint tension, reduce the risk of muscle-tendon injuries and return to resting state
Organization: Individual

3.7 Return to Resting State

Exercise #2

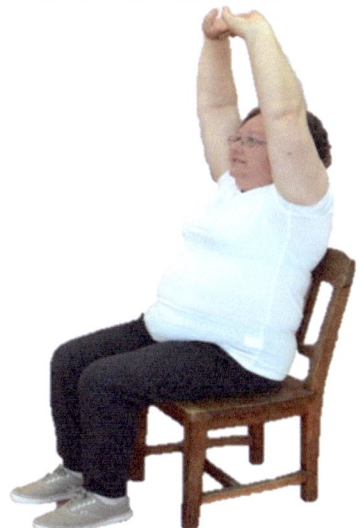

Interlace the fingers, point the palms facing upwards and stretch the arms

Objective: Reduce muscle and joint tension, reduce the risk of muscle-tendon injuries and return to resting state

Organization: Individual

Exercise #3

Interlace the fingers, point the palms facing upwards, and laterally stretch the arms to one side (switch sides at each exercise)

Objective: Reduce muscle and joint tension, reduce the risk of muscle-tendon injuries and return to resting state

Organization: Individual

Exercise #4

Hold the shoulder with the hand and, with the help of the opposite arm, raise the elbow (switch sides at each exercise)

Objective: Reduce muscle and joint tension, reduce the risk of muscle-tendon injuries and return to resting state

Organization: Individual

Exercise #5

Stretch one arm forward and press it down by holding the four fingers of the hand with the opposite hand (switch sides at each exercise)

Objective: Reduce muscle and joint tension, reduce the risk of muscle-tendon injuries and return to resting state

Organization: Individual

3.7 Return to Resting State

Exercise #6

Stretch one arm sideways and press it towards your chest with the opposite hand (switch sides at each exercise)

Objective: Reduce muscle and joint tension, reduce the risk of muscle-tendon injuries and return to resting state

Organization: Individual

Exercise #7

(continued)

(continued)

While standing up, bend one of the legs forward and extend the other one backwards, with the heel resting on the ground (switch sides at each exercise)
For seniors who have ambulatory capacity, they may remain seated and stretch the legs forward
Objective: Reduce muscle and joint tension, reduce the risk of muscle-tendon injuries and return to resting state
Organization: Individual

3.8 Conclusions and Practical Implications

The benefits of pursuing regular physical exercises in the treatment and control of degenerative diseases, such as diabetes, osteoporosis and cardiovascular diseases, are evident. It is, therefore, important to measure the musculoskeletal, cardio-respiratory and neurological abilities of the elderly, taking into account the evaluation of multiple physical parameters, such as cardio-respiratory capacity, muscular strength, flexibility, agility, body composition, among others.

Despite the thoroughly described activity programme, one further highlights that physical activity for the elderly should be provided by qualified professionals. Moreover, before starting any physical activity programme for this population, it is recommended that all participants undergo a rigorous and thorough medical evaluation. Put it differently, the prescription of a physical activity programme for the elderly entails knowing his/her limitations, existing pathologies and individual changes arising from ageing.

References

ACSM'S. (2000). *Guidelines for exercise testing and prescription*. Baltimore, MD: Lippincott Williams & Wilkins.
Baptista, F., & Sardinha, L. (2005). *Avaliação da actividade física e do equilíbrio de pessoas idosas: Baterias de Fullerton*. FMH Edições.
Cassilhas, R. C., Viana, V. A. R., Grassmann, V., Santos, R. T., Santos, R. F., Tufik, S., et al. (2007). The impact of exercise on the cognitive function of the elderly. *Medicine and Science in Sports and Exercise, 39*, 1401–1407.
Clark, J., & Cotton, R. T. (1998). *American council on exercise, exercise for older adults ace's guide for fitness professionals*. Champaign, IL: Human Kinetics Publishers.
Dias, G., & Mendes, R. (2013). *Atividade física para a terceira idade*. Escola Superior de Educação de Coimbra (ESE.IPC).
Dias, G., Mendes, P., & Mendes, R. (2013). *Actividade física na terceira idade*. III Congresso Internacional de Gerontologia Social (3CIGS), Coimbra, 15 de Maio de 2013.
Dias, G., Mendes, R., Serra, S. P., & Braquinho, A. (2014). *Envelhecimento Activo e Actividade Física*. Escola Superior de Educação de Coimbra: Coimbra.

References

Dias, G., Mendes, R., Serra e Silva, P., & Banquinho, M. A. (2015). *Gerontomotricidade: actividades lúdicas e pedagógicas para o corpo envelhecido* Eds: Dias G, Rui Mendes, Polybio Serra e Silva, Maria Aurora Banquinho. Coimbra. Escola Superior de Educação de Coimbra.

Geis, P. P. (2002). *Atividade física e saúde na terceira idade – teoria e prática*. Porto Alegre: Artmed.

Geis, P. P., & Rubi, M. C. (2003). *Terceira idade: actividades criativas e recursos práticos*. Porto Alegre: Artmed.

Hooke, A. P., & Zoller, M. B. (1992). *Active older adults in the YMCA: A resource manual*. Champaign, IL: Human Kinetics.

Irvine, A. B., Gelatt, V. A., Seeley, J. R., Macfarlane, P., & Gau, J. M. (2013). Web-based Intervention to promote physical activity by sedentary older adults: Randomized controlled trial. *Journal of Medical Internet Research, 15*(2), 1–19.

Jacob, L. (2007). *Animação de Idosos*. Porto: Ambar.

Jones, C. J., & Rose, D. J. (2005). *Physical activity instruction of older adults*. Champaign, IL: Human Kinetics Publishers.

Martins, R. A., Jones, J. G., Cumming, S. P., Coelho e Silva, M. J., Teixeira, A. M., & Veríssimo, M. T. (2012). Glycated hemoglobin and associated risk factors in older adults. *Cardiovascular Diabetology, 11*, 13.

Matsudo, S. M., & Matsudo, V. (1992). Prescrição de exercícios e benefícios da atividade física na terceira idade. *Revista Brasileira de Ciências e Movimento, 5*(4), 19–30.

Mcardle, W. D., Kacth, F. I., & Kacth, V. L. (2003). *Fisiologia do exercício: energia, nutrição e desempenho humano*. Rio de Janeiro: Guanabara Koogan.

Meirelles, M. A. E. (2000). *Atividade física na terceira idade*. Rio de Janeiro: Sprint.

Mendes, P., Dias, G., Almeida, M., Apóstolo, J., Mendes, R., & Panão, A. (2013). *Abordagem transdisciplinar na prática corporal do idoso*. III Congresso Internacional de Gerontologia Social (3CIGS), Coimbra, 15 de Maio de 2013.

Norman, V. K. (2010). *Exercise and wellness for older adults*. Champaign, IL: Human Kinetics Publishers.

Rikli, R., & Jones, C. J. (2001). Senior Fitness Test Manual. *Champaign, IL: Human Kinetics*.

Silva, P. S. (1993). *Era uma vez um coração*. Fundação Portuguesa de Cardiologia.

Singh, M. A. F. (2002). Exercise comes of age: Rationale and recommendations for a geriatric exercise prescription. *Journal of Gerontology Series A: Biological Sciences and Medical Sciences, 57*(5), 262–282.

Smith, J. C., Nielson, K. A., Woodard, J. L., Seidenberg, M., & Rao, S. M. (2013). Physical activity and brain function in older adults at increased risk for Alzheimer's. *Disease Brain Science, 3*(1), 54–83.

Spirduso, W. W. (2005). *Physical dimensions of aging*. Champaign: Human Kinetics.

Topp, R., Mikesky, A., Wigglesworth, J., & Holt, W. (1993). The effect of a 12-week dynamic resistance strength training program on gait velocity and balance of older adults. *Gerontologist, 33*(4), 501–506.

Chapter 4
Technology for the Active Senior

Micael Santos Couceiro and Gonçalo Nuno Figueiredo Dias

Abstract The ever-increasing ageing of the population has been under the international spotlight for the past few years. The solutions provided so far essentially focus on the health care, personal care services and pensions. However, even by reducing the growth of healthcare costs, the demographic trends and constrained state will drive health and retirement spending toward an even larger share of the economy. Therefore, one needs to go beyond the classical approaches and develop new breakthroughs for active and assisted living (AAL) by benefiting from the Information and Communications Technology (ICT). Those breakthroughs should focus on fighting the cognitive impairment, frailty and social exclusion of the ageing population, so as to improve the quality of their life and the ones surrounding them. The recent technological developments combined with other disciplines, such as behavioural, sociological, health and others, have opened a new window of opportunity. Nevertheless, and given the multidisciplinary research involved, the market has not yet provided reliable commercial solutions to appropriately address this specific challenge. This chapter outlines the general ideas and future prospects, supported by the recent rapid proliferation of ICT solutions, mainly focusing in two key approaches: (i) mixed reality serious games; and (ii) robotics. The main purpose is to foster the development of projects based on the premise that those will be ultimately beneficial, not only for the science and technological competitiveness, but also to our socio-economic welfare, increasing the sustainability and efficiency of both social and healthcare systems and, more importantly, the quality of life of the elderly.

Keywords Information and communication technology · Active and assisted living · Wearable technologies · Augmented reality · Robotics

4.1 Mixed Reality Serious Games and Robotics

The physical inactivity problem in the elderly population is of major concern across the globe. It represents a massive part of all financial and health issues faced by our society. Nevertheless, as shown in the literature, the most challenging problem resides in the lack of qualified human resources (e.g., physical education teachers) to meet the psychomotor needs of the ever-increasing elderly population (Dias and Mendes 2013). In that sense, the financial costs covering this particular type of care can be mitigated, in a long-term perspective, by introducing, for instance, service robotic solutions and mixed reality serious games, which can promote the physical activity at any time of the day and without any constraints associated to human carers. Associating these benefits to the cost, one can understand the added value of these technologies designed for the elderly care, having the main purpose of reducing expenses from the host institution and the senior citizen, decreasing the consumption of drugs for the heart, bones, cartilage, and other philological aspects inherent to the third and fourth age.

Service robotic solutions and mixed reality serious games play an important role in "bringing" physical activity programmes for this population in a personalized and specialized way. Although the assumptions underlying the operation of physical exercises in the institutionalized senior citizen are similar than to the one living in his/her own home, in the latter, the term "personal trainer" requires a more autonomous and adaptive meaning. For instance, the senior can be active and available for physical activity or other activities of social nature (e.g. spending time with family at home). Therefore, the technology needs to assume a strong "generational bond" with other humans without pre-existent information and be able to promote the benefits of physical activity by considering more generalized exercises (e.g. traditional games). This kind of inclusion through sports is of high relevance from both educational and pedagogical perspectives and finds virtues not only within the senior citizen but also within society in general. The impact of these solutions may be observed from different perspective (e.g. health, economic and social) and at different levels: (i) individual and institutional incomes; (ii) the decrease of individual and institutional outcomes; and (iii) the increase of healthy levels that directly and indirectly influence individual and institutional outcomes (e.g. medication, inpatient and outpatient costs related with current diseases, among others).

The institutional incomes are related with the overall improvement of the quality of life in the institutional recruitment and management of seniors and related services. In health and care institutions, an increase in seniors' health is reflected in fewer inpatient and outpatient service visits and medication, decreasing the demands around a given senior and, as such, the cost. In that perspective, the difference between a senior's monthly payment to one institution and the demands necessary to ensure his/her quality of life, is considerably larger and the economic institutional outcomes are inevitably more sustainable (Ageing Working Group 2012). So, individual incomes are related with health benefits, namely the development of a control system for monitoring human physiological parameters and the removal or reduction of medication by following an alternative therapy associated with physical activity (Ageing Working Group 2012).

4.1 Mixed Reality Serious Games and Robotics

The health care expenses are distributed over some public and private institutions. Public expenses are related to public tax and public social system, while the private expenses are related to all insurances covering individual diseases and work accidents (McCall et al. 2001). Also, a part of the expenses is engaged to the individual citizen. In 2010, public insurance coverage represents the major slice, with a sum of 67% of total expenditures, followed by 22% for out-of-pocket and 10% for private insurance. The total amount is approximately of about 150 billion of euros and, for 2020, the projection goes up to 190 billion of euros. In England, Wittenberg et al. (1998) shows that the long-term expenditure would need to rise 148% between 1996 and 2031 to meet demographic pressures and allow for real rises in costs. This represents an overall increase from about 10 billion of euros in 1996 to 14 billion of euros in 2010 and 25 billion of euros in 2031. Nevertheless, the literature clearly shows that all this can be minimized with public physical activity strategies. To make this happen, public institutions could request, to all citizens who need health and care services, a physical activity frequency. In other words, public and protocoled institutions that promote physical activity programmes have to resort to complex tools to improve health and functional fitness levels for all the senior citizen population. The imbalance between elder care service wages and the cost of these technologies may be partially remedied by projected ratios to total population. Using current country-specific proportions of the population providing family care by age group and gender, rough projections on the availability of caregivers in the population can be estimated and compared relatively to the expected number of dependent individuals (Barbesh and Glied 2010).

On the other hand, growth and technological development becomes ambivalent with regard to both physical and psychological health. Studies show that constant growth, associated with lifestyle deviations in general, and a significant increase in sedentary behaviour in particular, have led to the emergence of a larger number of disabled people, victims of traffic accidents, labour accidents, and cardiovascular events. Given the increase in the number of people with disabilities across Europe, due to population growth, ageing, emergence of chronic diseases, and medical advances that prolong life, it is necessary to preserve the functional independence of patients, while helping them to maintain their quality of life. So, the promotion of physical, psychological and social activities can be considered as a key-supporting element to overcome the socio-economic challenges facing people suffering from impairments or disabilities (Sallis and Owen 1997; Humpel et al. 2002).

Next sections further explore mixed reality serious games and service robotics designed for the active senior.

4.2 Mixed Reality Serious Games

Mixed reality serious games play a crucial role as a pragmatically viable tool in several aspects of teletherapy and telerehabilitation (Ma et al. 2014). Such games have been recognized as increasing patients' motivation in performance of intensive training tasks that may be repetitive and boring, as well as distracting individuals to aid in the management of pain (Callejas-Cuervo et al. 2015). This implementation aims to boost efforts in health promotion and prevention, with a focus on reducing health inequalities.

Bearing these ideas in mind, the authors suggest that one should support the scientific and technical development of applied gaming and gamification for rehabilitation purposes, not only to create novel ICT-based solutions and methods to address societal challenges facing people suffering from impairments and disabilities, but also to assist SMEs in seizing new business opportunities based on the new technology designed for therapeutic, recreational or rehabilitation purposes. The aim should be to increase the take up of gaming technologies within non-leisure contexts, by main-streaming mixed reality, online serious games, and interface technologies for social and economic benefits. Put it different, new R&D projects focusing in the design of mixed reality serious gaming intelligent solutions to provide physical and cognitive rehabilitation are needed. To do so, these solutions should consider multiple inputs besides multimodal human–machine interaction (HMI), such as the kinematic and physiological data of the elderly, by benefiting from multisensor fusion and classification for contextual information assessment and Internet-connected cloud services for real-time computing and communication with healthcare entities.

The connection between mixed reality and psychology has recently been strengthened, since it can be applied to exposure therapy. In this field, Rizzo et al. (2006) proposed virtual warfare scenarios to be used therapeutically with soldiers with post-war trauma. Another example are the studies conducted by Botella et al. (2010) and Rus-Calafell et al. (2013) that considered virtual reality for the treatment of psychological disorders. The success of these immersive systems depends on the sense of presence level that they can induce in a participant taking part in such experiences (Steuer 1992). The sense of presence is defined as the psychological sense of "being there", in the virtual environment or in an extension between both real and virtual environments. In other words, it is an emergent property based on the immersion given by the technology. For instance, people experiencing virtual reality may have the illusion of being in a virtual place and, consequently, carry out actions as if the situation and events depicted were really happening. Accompanying these actions, one may observe physiological changes, such as higher heart rates, reflexive behaviours, such as smiling at a virtual human character, and emotions such as the fear of dying in the virtual world (Reiners et al. 2014). For instance, recent studies found that a fake body part can be incorporated into human body representation through synchronous multisensory stimulation on the fake and corresponding real body part, such as the well-known "rubber hand illusion" (Sanchez-Vives and Slater 2005; Botvinick and Cohen 1998; Kilteni et al. 2012).

Even though the literature presents many works on mixed reality and others on serious games, adequately merging both is still a considerable unsolved challenge.

New mixed reality serious games should revolve around three key technologies: (i) serious games; (ii) mixed reality; and (iii) wearable technology.

4.2.1 Serious Games

Serious games are an emerging technology growing in importance for specialized training, especially for health, taking advantage of 3D games and game engines in order to improve the realistic experience of users. Many serious games for health have been proposed during the last few years. It is noteworthy, however, that such solutions are not intended as a substitute human carer, as it those can never substitute genuine human interactions between elderly and skilled clinical staff. Rather, these solutions should be designed to provide support, complement skilled clinical interventions, and offer help, while promoting health through rehabilitation, when family, friends or care providers are not available.

Previous research (e.g. Ma et al. 2014) has indicated that the use of serious games in this domain enables the personalization of a therapeutic regimen, based on each patient's background (e.g. interests, hobbies, skills, experiences, etc.) and psychological, physical and social needs (Laut et al. 2015). This user-centred design of therapeutic regimens could enhance each patient's motivation levels, engagement and adherence to prescribed rehabilitation programmes, thus facilitating the likelihood of better health outcomes (Li et al. 2014).

According to Wiemeyer and Kliem (2011) the use of serious games for the elderly has allowed specialists to determine the patient performance in terms of motor functions. However, the potential of use these capabilities to accurately perform medical assessment as a tool to guarantee effectiveness is not being exploited. From this perspective, a study by De Carvalho and Ishitani (2012) revealed that by playing serious games, the elderly can become more integrated into society, reducing the digital exclusion. Furthermore, according to Lemus-Zúñiga et al. (2015), information technology and serious games allow older population to remain independent for longer. Hence, when designing technology for this population, developmental changes, such as attention and/or perception, should be considered. For that purpose, mixed reality is a good option for targeting serious games.

4.2.2 Mixed Reality

Rehabilitation therapies involve regular and repetitive exercises or drills. Mundane rehabilitation drills can easily become routine and lead to boredom, decreasing patient motivation and engagement (Li et al. 2014). As a solution to this problem, on-going research on immersive virtual, augmented, or, more generally, mixed reality environments has shown their applicability in the physical rehabilitation and psychological domains (Laut et al. 2015). A body of work, such as Botella et al. (2010) and Lehrer et al. (2011), has identified significant benefits of using mixed reality serious games, defined as games that do not have entertainment, enjoyment or fun as their primary purpose (Djaouti et al. 2011). Rather, these activities have the intent to improve the quality of life of people with physical and psychological disabilities, either due to trauma or disease.

In a technological perspective, however, mixed reality serious games can be challenging, especially for such old population. In a technological perspective, there is the need of a head-mounted display (HMD) providing an immersive experience with mixed reality (includes sound feedback). The basic idea is to avoid the use of non-wearable devices, such as smartphones and tablets, so they can perform the desired actions without additional constraints (e.g. physical exercise). Such solution should be designed to help caregivers work more efficiently and effectively in attending to a larger elderly population in the same amount of time. The use of serious games (Ma et al. 2014), with the addition of mixed reality, enables the personalization of a therapeutic regimen, based on each patient's background (e.g. interests, hobbies, skills, experiences, etc.) and psychological, physical and social needs (Li et al. 2014; Laut et al. 2015). This user-centred design of therapeutic regimens could enhance each senior's motivation levels, engagement and adherence to prescribed rehabilitation programmes, thus facilitating the likelihood of better health outcomes (Calderita et al. 2013). Interactivity and continuous feedback provided to the patient creates immersive experiences during rehabilitation exercises that take place in mixed reality environments, thus keeping him/her continuously engaged. For instance, using an adaptive learning paradigm, the ICT-based system can automatically move to the next game level if a patient successfully completes a current level.

4.2.3 Wearable Technology

Besides the HMD, to interact with mixed reality serious games, one needs to develop a completely wearable set of small sensors (accelerometers, electromyography, electrocardiogram) and haptic actuators (force feedback) placed in the elderly's body through an intelligent suit, easy to put on and not requiring any sort of external sensors (e.g. a network camera). This should be integrated into an architecture for body motion analysis and biosignals data assessment to understand the elderly's emotional and psychosocial factors. This should have the intent to develop an innovative solution that allows the creation and manipulation of responsive mixed reality serious games, oriented to psychological and physical rehabilitation therapies, with improved features provided through the concepts of cloud computing and IoT.

The wearable technology comprises conductive fabrics and threads, as well as other flexible and/or lightweight components. This technology, combined with a HMD to provide an immersive mixed reality experience, allows to merge digital content with the surroundings. The virtual representation of the surroundings, on the other hand, can be defined by the team of carers. The main purpose is to not only perceive an environment mixing both real and virtual worlds, but to "feel" the virtual objects as if they were real. Such feeling can be promoted by the haptic feedback from the wearable technology, such as the feeling of the foot touching the floor, and the HMD, such as the sound of the foot touching the floor.

4.3 Robotics

It should be noted that the senior population, with trifling daily experiences and limited financial resources, rarely or never had contact with ground-breaking technologies, from which robotics probably takes the lead. The recent rapid proliferation of ICT devices, as service robotics, in everyday life can be rather unclear to the elderly population, especially because of the lack of past experience on which they can base an evolutionary understanding of the role of these new technologies in their lives (Hough 2011). Researchers have observed that negative emotional reactions manifest as embarrassment among the senior citizen, which may translate into hesitation to engage with new ICT (Heerink et al. 2008).

On the other hand, positive experiences have been shown to ease the acceptance of such technology and reduce loneliness and depression. The elderly has also demonstrated a positive emotional affinity for ICT devices with familiar form factors that resemble known objects (Wada and Shibata 2007). For instance, Tamura et al. (2004) compared the reactions of nursing home patients to an artificial fur covered toy dog that could wag its tail and sit, with Sony's AIBO, a well-known socially assistive robot dog with a smooth plastic body, that recognized up to 75 spoken commands and communicated via touch, sight, hearing and balance.

Remarkably, the multiple trials revealed that patients averaged approximately 225% more interactions with the toy dog than with AIBO, despite the robot's high level of interaction capability. Subsequent experiments also revealed that patient interaction levels with AIBO increased significantly when they identified it as a dog, either through intervention and explanation by an occupational therapist or by clothing AIBO in artificial fur similar to that of the toy dog. This is in line with the work of Kamei et al. (2012), stating that robots must acquire new features, closer to the real human needs, especially in what concerns the senior population, so as to create new quality products aimed at this market segment.

The authors in Kamei et al. (2012) also claim that robots have to go further than just being built around the three basic functions mainly used in machines, i.e. sensing, actuation and control. In the case of cooperation with the elderly in the context of physical activity, a particular case of human–robot interaction (HRI), the robot has to overwhelm these three functions, by presenting itself as an intelligent kinetic model similar to that offered by human beings in terms of physical exercise credibility (e.g. physical education teacher, personal trainer).

It is based on these considerations, and the limitations inherent to the state of the art and solutions available in the market, that new solutions should be proposed, mainly focusing in three key features: (i) appearance and physical characteristics; (ii) real-time assistance and monitoring over the Internet; and (iii) autonomous navigation and operation under dynamic environments.

4.3.1 Appearance and Physical Characteristics

The authors believe that and elderly centred solution would consist of a mobile robotic platform with humanoid appearance. The platform should not only be intended to fulfil most of the common tasks associated with social platforms, namely to act as a companion, but to primarily act as a robotic "personal trainer" that may promote healthy activity among senior citizens, thus consequently striving for a better quality of life. Therefore, such service robot should contribute to the prevention/treatment of osteoarticular diseases, or arthritis, as well as cardiovascular diseases, such as myocardial infarction while, at the same time, significantly reduce the costs related with healthcare and ADL assistance on active senior citizens.

To do so, the platform should then be designed in such a way to replicate the human movement, mainly focusing on completing exercises specifically designed for the elderly. This is accomplished by benefiting from a kinetic structure comprising of upper and lower limbs, as well as torso and neck, optimized based on the DoF necessary to ensure the needs of the senior population. Based on a preliminary evaluation carried out by the authors, the structure is expected to comprise 41 DoF, without considering the DoF associated to facial expressions. This number of DoF is adapted to the senior citizen maximum mobility capabilities and physical limitations. Depending on the target senior citizen, the robot should then be programmed to constrain its own exercises (i.e. reduce the number of DoF or the maximum amplitude of a given gesture). For instance, a senior citizen with 50% physical limitations that cannot move its lower limbs will not be subject to an activity profile that may include those.

Operationally, through various physical activity programmes easily programmable by carers through visual data mimicking, the service robot should be able to provide responsive exercises with upper and/or lower limbs (i.e. separately or simultaneously), flexion and extension of the torso, as well as the mobility of the neck, controlling both position and velocity of each DoF with high accuracy and precision. In practical terms, the seniors have to replicate, or mimic, the motions proposed by the service robot, according to previously established programmes. These programmes can be designed for a specific senior or target group (i.e. group-based exercise programmes), which are adapted based on multivariable identification, including face recognition and geo-positional data.

4.3.2 Real-Time Assistance and Monitoring Over the Internet

Given the above, the overall idea should be to develop a service robot that teaches the senior citizen how to perform adequate physical activities independently, thereby optimizing their psychomotor autonomy. Nevertheless, such activities require to be well-defined based on the target senior citizen, or group of senior citizens, by specialists, namely doctors or physical education teachers. Moreover, the service robot is responsive and adapts its kinetic structure and behaviour to the real-time contextual data retrieved. To that end, the robot should be equipped with the ability to passively monitor multiple vital and environmental parameters, such as the heart rate, body and room temperatures, odour's recognition, gas detection, among others.

All input data is processed and classified in real-time, using fuzzy-logic and Bayesian reasoning (Ferreira et al. 2013), by also considering the available information in the cloud, both provided by family or physicians, as well as updated by service robots themselves, following the insights of *RoboEarth* and other IoT robotic initiatives (Tenorth et al. 2012). The classification output of the monitoring system can be revealed upon a large set of different actions, namely emitting "voice" warnings or directly calling local emergency and urgent care services, depending on the irregularity of the readings and based on the insights (i.e. rules) provided by specialists. To ensure the feasibility of this mechanism, the service robot should be endowed with a pervasive access to the internet, by benefiting from 3G and 4G technologies as a fault-tolerant mechanism to local wireless infrastructures.

To ensure that the solution is fully autonomous, all computation should occur directly on the robot, without the unconditional need of an external infrastructure, communication, or human interaction beyond high-level commands. However, the existing infrastructure (3G and 4G technologies) should be used to exchange data and maintain updated mission information, such as the most up-to-date map the robot is operating, robot's position in real-time, important environmental and vital conditions, among others. Besides remote monitoring as stated, it is also important to highlight that the platform should also possess semi-autonomous and fully teleoperated modes. The robot can relegate important decisions to remote human operators when running in semi-autonomous mode, or when the assessed context through sensor data does not ensure properly informed decisions in the field (e.g. fire outbreak). This is possible by sending camera feeds via the communication infrastructure, and using an intuitive control interface suitable for the end-users.

In order to achieve a higher level of hardware abstraction, the control architecture should be hierarchically divided, and the Robot Operating System (ROS) should be adopted as the main framework in all platforms. ROS is currently the most popular robotic framework, being the closest one to become a de facto standard in Robotics (Quigley et al. 2009). Besides hardware abstraction and low-level device control, it provides an established implementation of commonly used functionalities in Robotics, and is ideal for the development of new algorithms, such as those referred to in this proposal. Similarly to other works, such as Arrichiello et al. (2006), this control loop organization allows using the same high-level algorithms with different types of robotic systems.

Considering the physical properties of such service platforms, typical social robotic tasks can also be ensured. For instance, the delivery of specific lightweight objects (e.g. medication, water glass), keeping a record of medication schedules, acting as a voice-dialling enabled service to call family, friends and carers, or being completely teleoperated over the Internet, are some of the most trivial low-level objectives these robots should provide. Moreover, given the dynamic nature of all these tasks (e.g. introduction of new medications and schedulers or addition of new telephone contacts), the data can be managed by the responsible entities and added to the cloud, without the need to reprogram the robot.

4.3.3 Autonomous Navigation and Operation Under Dynamic Environments

The physical properties of these robots should make it possible to autonomously navigate in multi-floor sites (e.g. healthcare institutes or personal homes), thus following the trends presented around legged robots as a solution to cope with stairwells constraints, such as Cupec et al. (2003) and Gutmann et al. (2004) works. The fulfilment of these type of tasks, on the other hand, can benefit from defining, a priori, the physical 3D structure wherein the robot will operate. Nevertheless, although feasible for robots operating in health and care institutes (in which experts are in charge of the surrounding environment), it is hard, or almost impossible, to ensure that personal homes (seniors' residences) will maintain the same physical properties over time (e.g. addition of new furniture and/or moving a given furniture from one place to another). In spite of this, even though state-of-the-art probabilistic techniques in known maps can be used (e.g. adaptive Monte Carlo localization), given the dynamic nature of these applications, the platforms should also benefit from long-term Graph-based simultaneous localization and mapping (SLAM) strategies (Walcott-Bryant et al. 2012).

The algorithm should be designed to enable the robot's localization even under an environment that may suffer changes over time. The SLAM state estimation engine can also benefit from incremental smoothing (Kaess et al. 2008) and fractional calculus memory effect (Couceiro et al. 2012). This allows robots to identify and revisit previously mapped areas even under minor changes. The visual depth SLAM approach can be fed with time-of-flight (ToF) camera information, thus benefiting from 3D scan matching to derive constraints for SLAM state estimation (Borrmann et al. 2008). The detection of changes can be considered by dynamically clustering the 3D scans for the same portion of the environment at different times. The persistency of such changes in a given location will update the map, and remove from the dynamic graph old poses and scans that no longer match the current state of the world.

Besides monitoring and assisting abilities, the robot energy autonomy is a major concern in these applications. Mobile robots' energy autonomy has benefitted from recent technological advances over the past few years, namely with the discovery of new material (e.g. solvents, polymers, etc.) (Cohen et al. 2008). Recently, Lithium-ion polymer (LiPo) batteries have been used for most lightweight mobile platforms with long-term autonomy (e.g. UAVs). This technology fostered the design of new solutions to interact increasingly with human environments, working with and around humans on a daily basis. However, under the current case study, i.e. interacting alongside unexperienced users, such as senior citizens, the mobile platform cannot be endowed with LiPo batteries since those are susceptible to explosion or fire. In order to overcome this drawback, this kind of mobile robot must exhibit some form of self-sustainability.

In addition to being robust in its physical design and control architecture as previously described, such robot must be capable of long-term autonomy (Silverman et al. 2002). As such, it is the authors' opinion that this type of robot should be equipped with state-of-the-art lithium iron phosphate (LiFePO$_4$) batteries offering advantages over other lithium chemistries in terms of thermal and chemical stability, thus improving battery safety. LiFePO$_4$ batteries have the lifetime of approximately three times longer than that of standard lithium-family or nickel-family batteries. These batteries have been recently used in electric vehicles and other platforms, such as *QUICC*,[1] *KillaCycle*[2] and *ZBoard*.[3]

Moreover, a custom recharging station can be designed to easily support the platform. The station should be on a predefined location known by the robotic platform and can comprise of a chair-type docking structure to support omnidirectional recharging. Under low battery status, the robot will simply sit in the recharging chair, thus physically docking to recharge batteries. These features allow overcoming the traditional need of a perfect relative pose between the robot and the docking station (Wu et al. 2009). It is noteworthy that, even while charging, the platform should remain with elementary features available, such as real-time monitoring.

4.4 Conclusions and Practical Implications

Service robotic solutions and mixed reality serious games can promote the necessary treatment and rehabilitation of various diseases, such as Parkinson's and Alzheimer's disease, among others, assuming a key role in optimizing the quality of

---

[1] www.quicc.eu.
[2] www.killacycle.com.
[3] www.zboardshop.com.

life of the elderly. This is a simple approach of healthy mind and body, where health professionals, physiotherapy and sport scientists can have a relevant action through the use of these new technologies to the elderly. For that purpose, we face new methodologies that enhance the elderly closer contact with the surrounding reality and the social environment.

Note that new technologies abruptly evolved over the past few years and the scientific community had to follow this evolution and transfer all their knowledge for the benefit of an ageing community that will dominate in the modern world. In that sense, it is certain that we all, without exception, will have to learn to live with an older body. The big question is that, regardless of this inevitable fatality, the elderly may, through the use of service robotic solutions and mixed reality serious games, continue to live in a balanced way, smoothly and without routines.

There is, however, still much work ahead before this kind of approach is completely available for the elderly, not only because of the R&D costs involved, but also because of the natural difficulty of implementing and validating them in real operation context (e.g. Hospitals, Psychomotor Rehabilitation Centres, etc.). However, international collaborative projects have been emerging over the past few years, bringing academia closer to the industry and to the end-user. It is then foreseen that one might expect to interact with these technologies in a near future.

References

Ageing Working Group. (2012). *The 2012 Ageing Report-Economic and budgetary projections for the 27 EU Member States (2010–2060)*. Brussels: EU Commission.
Arrichiello, F., Chiaverini, S., & Fossen, T. I. (2006). Formation control of marine surface vessels using the Null-Space-based Behavioral control. In *Group Coordination and Cooperative Control* (pp. 1–19). Berlin: Springer.
Barbesh, G. I., & Glied, S. A. (2010). New technology and health care costs—The case of robot-assisted surgery. *The New England Journal of Medicine, 363*(8), 701–704.
Borrmann, D., Elseberg, J., Lingemann, K., Nüchter, A., & Hertzberg, J. (2008). Globally consistent 3D mapping with scan matching. *Robotics and Autonomous Systems, 56*(2), 130–142.
Botella, C., Bretón-López, J., Quero, S., Baños, R., & García-Palacios, A. (2010). Treating cockroach phobia with augmented reality. *Behavior Therapy, 41*(3), 401–413.
Botvinick, M., & Cohen, J. (1998). Rubber hands 'feel' touch that eyes see. *Nature, 39*(6669), 756.
Calderita, L. V., Bustos, P., Mejías, C. S., González, B. F., & Bandera, A. (2013). Rehabilitation for children while playing with a robotic assistant in a serious game. In *Neurotechnix 2013 Proceedings, Vilamoura, Algarve, Portugal*.
Callejas-Cuervo, M., Díaz, G. M., & Ruíz-Olaya, A. F. (2015). Integration of emerging motion capture technologies and videogames for human upper-limb telerehabilitation: A systematic review. *Dyna, 82*(189), 68–75.
Cohen, D. A., Ozick, D., Vu, C., Lynch, J., & Mass, P. R. (2008). *Autonomous robot auto-docking and energy management systems and methods*. Patent-7332890.
Couceiro, M. S., Martins, F. M., Rocha, R. P., Ferreira, N. M., & Sivasundaram, S. (2012). Introducing the fractional order robotic Darwinian PSO. In *AIP Conference Proceedings-American Institute of Physics* (Vol. 1493, No. 1, p. 242).

References

Cupec, R., Lorch, O., & Schmidt, G. (2003). Vision-guided humanoid walking—Concepts and experiments. *Autonome Mobile Systeme 2003* (pp. 1–11). Berlin: Springer.

De Carvalho, R., & Ishitani, L. (2012). Motivational factors for mobile serious games for elderly users. *Paper presented at the Proceedings of XI SB Games 2012*, Brasilia, Brazil.

Dias, G., & Mendes, R. (2013). *Atividade Física para a Terceira idade*. Coimbra.

Djaouti, D., Alvarez, J., Jessel, J. P., & Rampnoux, O. (2011). Origins of serious games. *Serious games and edutainment applications* (pp. 25–43). London: Springer.

Ferreira, N. L., Couceiro, M. S., Araújo, A., & Rocha, R. P. (2013). Multi-sensor fusion and classification with mobile robots for situation awareness in urban search and rescue using ROS. In *2013 IEEE International Symposium on Safety, Security, and Rescue Robotics (SSRR)* (pp. 1–6). IEEE.

Gutmann, J. S., Fukuchi, M., & Fujita, M. (2004). Stair climbing for humanoid robots using stereo vision. In *Intelligent Robots and Systems, 2004. (IROS 2004). Proceedings. 2004 IEEE/RSJ International Conference* (Vol. 2, pp. 1407–1413). IEEE.

Heerink, M., Kröse, B., Wielinga, B., & Evers, V. (2008). Enjoyment intention to use and actual use of a conversational robot by elderly people. *In Proceedings of the 3rd ACM/IEEE International Conference on Human Robot Interaction* (pp. 113–120). ACM.

Hough, M. G. (2011). Exploring elder consumers' interactions with information technology. *Journal of Business & Economics Research (JBER), 2*(6), 61–66.

Humpel, N., Owen, N., & Leslie, E. (2002). Environmental factors associated with adults' participation in physical activity: A review. *American Journal of Preventive Medicine, 22*(3), 188–199.

Kaess, M., Ranganathan, A., & Dellaert, F. (2008). iSAM: Incremental smoothing and mapping. *IEEE Transactions on Robotics, 24*(6), 1365–1378.

Kamei, K., Nishio, S., Hagita, N., & Sato, M. (2012). Cloud networked robotics. *IEEE Network, 26*(3), 28–34.

Kilteni, K., Normand, J. M., Sanchez-Vives, M. V., & Slater, M. (2012). Extending body space in immersive virtual reality: A very long arm illusion. *PloS one, 7*(7), e40867.

Laut, J., Cappa, F., Nov, O., & Porfiri, M. (2015). Increasing patient engagement in rehabilitation exercises using computer-based citizen science. *PLoS ONE, 10*(3), e0117013.

Lehrer, N., Chen, Y., Duff, M., Wolf, S. L., & Rikakis, T. (2011). Exploring the bases for a mixed reality stroke rehabilitation system, Part II: Design of interactive feedback for upper limb rehabilitation. *Journal of Neuroengineering and Rehabilitation, 8*(1), 54.

Lemus-Zúñiga, L. G., Navarro-Pardo, E., MoretTatay, C., & Pocinho, R. (2015). Serious games for elderly continuous monitoring. In *Data Mining in Clinical Medicine* (pp. 259–267). New York: Springer.

Li, C., Rusák, Z., Horváth, I., Hou, Y., & Ji, L. (2014). *Optimizing patients' engagement by a cyber-physical rehabilitation system*. In *GI-Jahrestagung* (pp. 1971–1976).

Ma, M., Jain, L. C., & Anderson, P. (Eds.). (2014). *Virtual, augmented reality and serious games for healthcare 1* (Vol. 1). Springer.

McCall, N., Komisar, H. L., Petersons, A., & Moore, S. (2001). Medicare home health before and after the BBA. *Health Affairs, 20*(3), 189–198.

Quigley, M., Conley, K., Gerkey, B., Faust, J., Foote, T., Leibs, J., et al. (2009). ROS: an open-source Robot Operating System. In *ICRA Workshop on Open Source Software* (Vol. 3, pp. 1–6).

Reiners, T., Teras, H., Chang, V., Wood, L. C., Gregory, S., Gibson, D., et al. (2014). Authentic, immersive, and emotional experience in virtual learning environments: the fear of dying as an important learning experience in a simulation. Transformative, Innovative and Engaging. In *Proceedings of the 23rd Annual Teaching Learning Forum* (pp. 1–14). University of Western Australia.

Rizzo, A., Pair, J., Graap, K., Manson, B., McNerney, P. J., Wiederhold, B., et al. (2006). A virtual reality exposure therapy application for Iraq War military personnel with post traumatic stress disorder: From training to toy to treatment. NATO Security through Science Series E. *Human and Societal Dynamics, 6*, 235.

Rus-Calafell, M., Gutiérrez-Maldonado, J., Botella, C., & Baños, R. M. (2013). Virtual reality exposure and imaginal exposure in the treatment of fear of flying: A pilot study. *Behavior Modification*, 0145445513482969.

Sallis, J., & Owen, N. (1997). Ecological models. In K. Glanz, B. K. Rimer, & F. M. Lewis (Eds.), *Health behavior and health education: theory, research and practice* (2nd ed., pp. 403–424). San Francisco, CA: Jossey-Bass.

Sanchez-Vives, M. V., & Slater, M. (2005). From presence to consciousness through virtual reality. *Nature Reviews Neuroscience, 6*(4), 332–339.

Steuer, J. (1992). Defining virtual reality: Dimensions determining telepresence. *Journal of Communication, 42*(4), 73–93.

Tamura, T., Yonemitsu, S., Itoh, A., Oikawa, D., Kwakami, A., Higashi, Y., et al. (2004). Is an Entertainment Robot Useful in the Care of Elderly People With Severe Dementia? *Journal of Gerentology: Medical Sciences, 59*(1), 83–85.

Tenorth, M., Perzylo, A. C., Lafrenz, R., & Beetz, M. (2012). The roboearth language: Representing and exchanging knowledge about actions, objects, and environments. In *2012 IEEE International Conference on Robotics and Automation (ICRA)* (pp. 1284–1289). IEEE.

Wada, K., & Shibata, T. (2007, October). Living with seal robots—It's sociopsychological and physiological influences on the elderly at a care house. IEEE *Transactions on Robotics, 23*(5), 972–980.

Walcott-Bryant, A., Kaess, M., Johannsson, H., & Leonard, J. J. (2012). Dynamic pose graph SLAM: Long-term mapping in low dynamic environments. In *2012 IEEE/RSJ International Conference on Intelligent Robots and Systems (IROS)* (pp. 1871–1878).

Wiemeyer, J., & Kliem, A. (2011). Serious games in prevention and rehabilitation—A new panacea for elderly people? *European Review of Aging and Physical Activity, 9*, 41–50.

Wittenberg, R., Pickard, L., Comas-Herrera, A., Davies, B., & Darton, R. (1998). *Demand for Long-term Care: Projections of Long-term Care Finance for Elderly People*. Canterbury: PSSRU, University of Kent. Available from the PSSRU website at www.ukc.ac.uk/PSSRU.

Wu, Y. C., Teng, M. C., & Tsai, Y. J. (2009). Robot docking station for automatic battery exchanging and charging. In *IEEE International Conference on Robotics and Biomimetics. ROBIO 2008* (pp. 1043–1046). IEEE.

Chapter 5
Conclusions

Gonçalo Nuno Figueiredo Dias and Micael Santos Couceiro

Abstract Narrowing down all that was previously presented to a sentence, the focus of this short book was to exploit the topic of active ageing and physical activity in a multidisciplinary perspective. This Springer Brief presented itself as a compendium that scientifically described the major pathologies and psychomotor difficulties plaguing the ageing body, presenting several exercises adequately represented with real pictures, which allowed to illustrate, in detail, the methodologies presented in the contemporary literature on the topic. This final chapter presents a short summary of the book, discussing its practical implications and providing some final recommendations for the active senior citizen.

Keywords Practical Implications · Recommendations · Ageing · Health · Active life

5.1 Conclusions

Active ageing means the maintenance of both motor and intellectual activities when they retire and move from an active to inactive state. Hence, it is urgent to implement healthy lifestyles and good health habits whose recommendations are well known in Western societies. Mostly everything surrounding the elderly defines active ageing, namely culture, gender, health systems, social services, behaviours, lifestyles, personal aspects, physical environment, social environment, and economic factors. Accordingly, it is necessary to optimize health, participation, and security opportunities, in order to improve the overall quality of life of the elderly.

Active ageing may contribute to reduce the morbidity and mortality in the older population, allowing to further reduce health care costs that usually increase as this population ages. Regular physical activity can improve the quality of life of the elderly and help controlling degenerative diseases (e.g., osteoporosis, diabetes, and cardiovascular diseases).

5.2 Practical Implications

Physical activity programmes, as described in this work, may have practical applications in the physical, mental, and social well-being of the elderly, simultaneously reducing the risk of chronic diseases, including hypertension, type-2 diabetes, osteoporosis, obesity, colon cancer, breast cancer, among other diseases.

However, a physical activity programme designed for seniors implies the knowledge of the physical limitations, existing pathologies, and individual changes arising from ageing. Therefore, before starting any physical activity programme for the elderly, participants should undergo a rigorous and thorough medical evaluation.

Likewise, it is also necessary to gather as much additional information (e.g., morphological characteristics, functional fitness, medical history, medication support, and daily routines).

5.3 Recommendations

As last remarks, daily 30–45-min walks are recommended for the elderly.

Strict eating habits and appetite control in the elderly is required, especially at the level of salt consumption. The elderly may opt for a Mediterranean diet, which provides a healthy and balanced nutrition.

Blood pressure should also be supervised in a daily basis by the elderly or health professionals.

A regular physical activity programme, as described in this work, can be complemented with other initiatives (e.g., yoga, walking, swimming, and cycling).

MIX
Papier aus verantwortungsvollen Quellen
Paper from responsible sources
FSC® C105338

If you have any concerns about our products,
you can contact us on
ProductSafety@springernature.com

In case Publisher is established outside the EU,
the EU authorized representative is:
Springer Nature Customer Service Center GmbH
Europaplatz 3, 69115 Heidelberg, Germany

Printed by Libri Plureos GmbH
in Hamburg, Germany